HOME WIRING

Rudolf F. Graf is an author whose name is familiar to engineers, technicians, do-it-yourselfers and hobbyists who have read some of his many books and hundreds of articles on mechanics, electricity, electronics and automotives. Mr. Graf is a graduate electrical engineer and received his MBA at New York University. He is also a licensed amateur radio operator and holder of a first-class radio telephone operator's license.

George J. Whalen is a writer, inventor and consultant specializing in physics, mechanics, electricity/electronics, and automotives, and he has been awarded several US patents in these fields. He is Director of Communications for a prominent medical electronics company. Mr. Whalen received his BA from Hofstra University. He, too, is a licensed radio amateur operator.

Graf and Whalen's how-to books and articles have a well-deserved reputation for making complex technology easily understandable and enjoyable. They have invented several useful mechanisms and hold a number of patents in the fields of thermodynamics, solid-state electronics, mechanics and chemistry. They have published hundreds of articles in leading technical publications as well as in popular do-it-yourself and how-to publications. They are well known authors and have more than a dozen engineering and technical books to their credit.

HOME WIRING

RUDOLF F. GRAF

GEORGE J. WHALEN

PRENTICE-HALL, INC., *Englewood Cliffs, New Jersey 07632*

Library of Congress Cataloging in Publication Data

GRAF, RUDOLF F.
Home wiring.

Includes index.
1. Electric wiring, Interior.   I. Whalen, George J.
II. Title.
TK3285.G68   621.319'24   81-7366
                                     AACR2
ISBN   0-13-392977-9

Editorial production supervision and interior
design by *Virginia Huebner*
Cover design by *Dawn L. Stanley*
Manufacturing buyer: *Joyce Levatino/Gordon Osbourne*

Printed in the United States of America

10   9   8   7   6   5   4   3   2

Prentice-Hall International, Inc., *London*
Prentice-Hall of Australia Pty. Limited, *Sydney*
Prentice-Hall of Canada, Ltd., *Toronto*
Prentice-Hall of India Private Limited, *New Delhi*
Prentice-Hall of Japan, Inc., *Tokyo*
Prentice-Hall of Southeast Asia Pte. Ltd., *Singapore*
Whitehall Books Limited, *Wellington, New Zealand*

CONTENTS

10 SPECIAL CIRCUITS: BELLS—FURNACE—SECURITY— INTERCOM—EMERGENCY GENERATOR 155

11 OUTDOOR WIRING: PORCH, YARD AND POOL LIGHTING LOW VOLTAGE SYSTEMS 169

12 TEST AND TROUBLESHOOTING—FLUORESCENT LAMPS, MOTORS, AND APPLIANCE REPAIR 179

PREFACE

Are you among the great number of homeowners who are intimidated by electricity? Are you convinced that it is so mysterious that you can't handle an electrical installation or repair? Do you feel doomed to seek professional help, with the attendant long delays for small jobs and astronomical costs for big jobs that may put a needed system improvement out of your reach? If your answer to any of these questions is "yes", this book is definitely for *you*!

The authors of this book are convinced that home electrical work *can be a cinch*! Electricity is not something to be taken lightly, but with a little enjoyable study and careful effort you can, in your spare time, unravel the mysteries, learn the techniques and *do-it-yourself*.

You might start by mapping your system, checking for inadequacies (particularly bad wire insulation), repairing outlets, improving switch circuits, etc. Later, you can add circuits for new appliances, and even substitute a new circuit breaker entrance panel to increase the capacity of your system. In a new house, with some careful work and self-training completed, you could even tackle the entire installation. Sound impossible? Emphatically no! Every day, people like you are finding that you *can* do a first-rate electrical wiring job, once you've got the basics straight. There is one small catch. For electrical work of any size, most municipalities require permits and inspections of the finished work. This is something you should want as much as the authorities do. There will be extra charges involved but they are so well worthwhile that you should never try to bypass these rules. They are designed to ensure your safety and to prevent fires. Therefore, after studying and planning your system, before buying the equipment, your first job should be to check all local rules and regulations, cover all necessary paperwork and legal steps to be sure you are proceeding properly.

This book will require you to do some skipping around when you get down to the actual work. The myriad possible designs and variations of typical electrical system installations make it impossible to cover every combination or to lay it out in one straightforward procedure. The best suggestion is to read through once very carefully, then back-track and pick and choose from among the sections to cover the repair, system add-on, or system installation that you have in mind.

Basically, this book covers two major topics—*what* and *how*. The first three chapters deal with electricity; what electrical codes require of the installer, and what safety measures should be taken; and finally the basics of system design and installation. You could say that this is the *electrical theory section*.

The remaining nine chapters are "how to" sections, focusing on the details of performing the tasks involved. Chapter 4 describes the electrical hardware, namely wires, fixtures and tools; Chapter 5 tells you how to bend, shape, form and connect wires and other hardware, using the appropriate tools. With that under your hat, how to install parts of the system is covered in Chapters 6 through 12 as follows: 6—Service Entrance Equipment, including the branch circuit connections, fuses and circuit breakers; 7 and 8—How to install outlets and switches on branch circuits; 9—Lighting; 10—Low voltage and other special equipment (furnace, emergency generators, etc.); and 11—Outdoor Wiring (lighting, pools, and such). Finally, Chapter 12 covers test and troubleshooting of the circuits and attached appliances as well as principles of appliances.

The authors of this book are especially indebted to Richard A. Shaw, whose many important contributions have helped make this book a reality, and to Mrs. John J. Dillon, for her truly heroic efforts at the typewriter. We would also like to express our thanks to the many firms, agencies, and sources, whose assistance brought each topic into sharper focus. All of the contributions are gratefully acknowledged.

And so, we present this book to you, in the hope that you will find that electrical wiring *can* be a cinch, and that you *can* do things you never thought possible, once you have the knowledge and confidence to try them.

RUDOLF F. GRAF

GEORGE J. WHALEN

New Rochelle, N. Y.

HOME WIRING

1

ELECTRICITY

What You Need To Know About It

If you look back over the long history of mankind, a few towering discoveries stand high above all of man's other achievements: *fire,* which helped early man to survive against cruel Nature's cold; *the wheel,* which gave man the ability to transport great burdens his frail muscles could never lift; *tools,* which gave man the ability to convert his hands into countless different forms to grip, pound, bend, twist, cut, shape, and manipulate every kind of object and material in his world. But one discovery stands taller than all others as a monument to man's genius, and that is the discovery of the incredible, invisible energy source we call *electricity.* Beginning with the first feeble twitchings of a frog's leg stimulated by a crude battery fashioned by Alessandro Volta in 1800, electricity has surged into our lives as the indispensable power source that has propelled our civilization into its present advanced state.

No greater proof of electricity's importance to our modern society is needed than its loss. Robbed of power by a storm or utility failure, we are suddenly hurled backward through time to an age of candle-lit nontechnology. Then, we can begin to understand what electricity truly means to us. It is warmth and light, the carefree acceptance of food that is preserved in refrigerators and freezers or toast that is precisely browned. It is clean clothes and dishes without back-breaking work. It is knowing the correct time and the events taking place in far corners of the world that shape our destiny. It is a home swept clean in minutes instead of hours. It is a cleanly drilled hole or a squarely cut timber. It is all of these and so much more that only when we lose it can we appreciate the fact that electricity is *life* as we know it today.

WHAT IS ELECTRICITY?

Most of us describe electricity in terms of its effects: light, heat, cold, turning-effort, and so on. That is because we can see or feel the results of these effects. But, by itself, electricity has none of these properties. Just as a tank of gasoline is nothing but a cold, smelly liquid until it is fed into an engine, so electricity is a form of energy that does *nothing* until it passes through the right kind of an energy converter, such as a light bulb, electric motor, heater, or other device. It is these converters that make electricity's energy useful to us; they convert electric power into equivalent forms of power that do something useful in the real world.

To really understand electricity, we must go down into the submicroscopic world of the *atom*, the tiny particles that make up the world that we can see with our own eyes.

In the world of the atom, every kind of substance we can imagine consists of a *nucleus* of *protons* around which we find rings of *electrons* orbiting the nucleus much as our earth orbits the sun. In simple atoms such as hydrogen, a single proton nucleus is positively-charged and a single negatively-charged electron orbits the nucleus at a distance that balances its negative charge against the positive charge of the nucleus (see Fig. 1-1). As substances grow in complexity, the number of electrons also grows. But the essence of the atom is that there are uniform, negatively-charged electrons orbiting and counter-balancing the charge of a positive, central nucleus. The number of electrons that can orbit the nucleus of any atom at a given distance is fixed: a maximum of two in the innermost ring, eight in the next distant ring, and so forth. An atom remains stable and identifiable as a specific kind of material because of this balance. The whole idea is that the sum of the negative charges of all the electrons equals the sum of the positive charges of the protons in the nucleus.

This might not sound very important, but let's connect this fact to a fundamental law of nature:

> *Particles with electric charges of the same sign (two negatives or two positives) will repel each other. Also, particles of opposite signs (a positive and a negative) will attract each other.*

Now, can you see why the structural balance of the atom is so delicate? The *distance* of the electrons from the nucleus determines the attraction force between them. This plays an important part in determining how easily one or more electrons can be pulled away from or pushed into an atom by applying an external charge.

Where do we get this external charge? Well, it may come from a *battery* or from a *generator*. Both of these are simply devices for creating a surplus of electrons at one terminal and a scarcity of electrons at a second terminal. Between the two terminals, we create a kind of electrical-charge pressure, which is called *voltage* [the unit of measure for voltage is the *volt* (V)]. A battery creates this

2

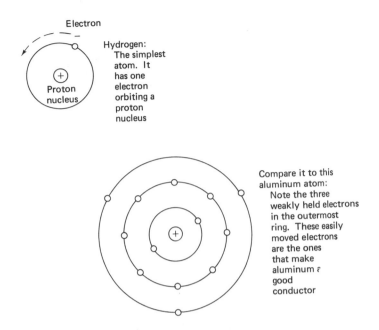

Fig. 1-1. A close-up look at the atom.

pressure through chemical action. A generator creates this pressure by converting mechanical effort into its equivalent electrical value. In both cases, a surplus of electrons results in one place and a corresponding scarcity of electrons at another place; that creates the voltage, and the device that causes this is called a *source*. These terms and processes are discussed in depth later.

CONDUCTORS AND INSULATORS

Electricity can only flow through materials called *conductors* (such as copper, aluminum, most other metals, as well as carbon and a host of others). The reason is that these materials are made up of atoms with one or more loosely bound electrons that can easily move from atom to atom if a voltage is applied.

On the other hand, some materials have atoms with such tightly-bound electrons (such as glass, mica, air, porcelain, rubber, many plastics and ceramics) that it is nearly impossible to get electrons to move from atom to atom. These materials are called *insulators*.

In electrical work, conductors (usually copper wire, brass parts, etc.) carry electrons from the source, to the load (e.g., a lamp or motor), and back to the source. Along the way, the conductors are surrounded by insulating material to prevent unwanted contact with the outside world. Without the proper use of conductors in combination with insulators, it would not be possible to generate or safely use electricity.

3

VOLTAGE

A conductor is just a piece of inert material until we connect it to a source of electricity. Once connected, the conductor's atoms are subjected to a pushing force on one side and a pulling force on the other. That force (or, as it was referred to earlier, electrical pressure) is known as voltage. In effect, it is the influence of the charges of all the electrons abundantly available at one terminal of the source relative to the scarcity of electrons at the other terminal.

Let's see how this works in an ordinary flashlight cell. The cell has two ends or terminals indicated as *positive* (+), and negative (−). By a chemical reaction within the cell, a surplus of electrons (all of which, you will recall, have a negative charge) are piled up at the negative terminal of the cell. The same chemical reaction in the cell produces a corresponding scarcity of electrons at the positive terminal. Atoms of the material connected to the positive terminal almost "yearn" to be fulfilled by the abundance of electrons at the negative terminal. You might think of this as a pushing force at the negative terminal and an equivalent pulling force at the positive. The quantitative difference in force is the voltage of the cell. It is a measure of the cell's potential capacity to do work.

Voltage gets its name from Alessandro Volta (1745-1827), an early physicist of Italy who discovered the chemical reaction between dissimilar metals that produced electricity. In short, he invented the electric cell. By stacking one cell atop another, he discovered that the electric force could be increased. And so, he also created the battery.

Voltage exists between the terminals of a cell or battery whether or not the cell is being used. In much the same way, voltage exists between the two contacts of every household alternating-current outlet in your home. It's there, whether or not some device is being powered. [Of course, a cell produces *direct current (dc)*, by chemical reaction; the outlet in your home provides *alternating current (ac)*, generated by huge alternators at the utility company's plant. These two forms of electricity differ in several ways, but the essential principles are the same.]

CURRENT

How do we make use of the voltage of a source? Well, we simply connect a suitable kind of conductor between the two terminals of the source. Now, the electrons at one terminal can reach the electron-hungry atoms at the other terminal. This movement of the electrons through the conductor is called *current*.

Going back to our flashing cell, let's connect a flashlight lamp between its terminals. When contact is made, electrons pour from the negative terminal of the cell, race through the filament of the lamp, and end at the positive terminal, filling some of the empty electron spaces in the atoms of the cell's positive terminal material. The current (movement of electrons) is a direct result of the voltage. When current flows, electrons are pushed into one end of the lamp filament, displacing the loose outer electrons from some of the filament's atoms. These loose electrons, in turn, displace the outer electrons of the next atoms, and so forth, all the way to the other end of the filament. This occurs almost instantaneously and is somewhat like pushing a billiard ball into a smooth pipe full of billiard balls. The instant one is pushed in, another pops out the other end. The effect is felt instantaneously; all balls shift together, but the ball you pushed in has not gone very far. In fact, it doesn't move unless you push more balls into the pipe. Pushing those balls or electrons in the previous example is the job of the electrical source (see Fig. 1–2).

Though the electric effect is felt throughout the entire filament instantly, the electrons migrate through the filament at a finite rate. The speed is determined by how many electrons are pushed into the filament, which is approximately a few inches per second with only one ampere of current flow.

Ampere, perhaps a new word to you, is simply a current measurement unit expressing *how many electrons are flowing past a given point each second*. An ampere (A) is 6.28×10^{18} (6,280,000,000,000,000,000) electrons flowing past a point in one second. In other words, with one ampere of current flowing, this huge number of electrons moves into one end of the circuit, the same number of electrons are bumped or pulled out the other end, while all intermediate electrons move just a small distance.

Ampere, as a measurement unit, is named in honor of the French physicist Andre Marie Ampere (1775–1836). His early experiments explained how current flows in a conductor and related that flow to electromagnetism, which we discuss later.

RESISTANCE

All conductors are not identical. Some, like copper, allow current to flow with ease. In other words, these conductors have a *low electrical resistance*. Tungsten, on the other hand, has a more tightly-bound atomic structure than copper. That

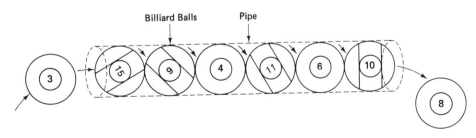

Fig. 1-2. As in a pipe full of billiard balls shown above, electron flow is felt instantly even through miles of conductor. Pushing an electron in one end will push an electron out the other at the same time.

is, it has a *higher resistance* to current than does copper. And so, when current flows in tungsten, the moving electrons must "squeeze" through its tight atomic structure. Collisions and friction occur and produce heating of the tungsten atoms. The result? Well, in the case of a lamp filament, the frictional heating causes the filament to glow, or *incandesce,* so that it gives off both heat and light. This is a form of power conversion that we see and use every day.

Although resistance is a benefit in devices like lamps, heaters, and cooking appliances, it is a nuisance in an ordinary run of conductor that is used to convey electricity. The reason is that resistance acts like a "kink" in a garden hose. It opposes the flow of electrons and thus limits the number that can appear at the conductor ends, where voltage is needed to push current through a load (such as a lamp, motor, heater, etc.). Resistance also results in wasteful heating of the conductor, which converts some of the potential energy available into heat energy.

Resistance can be introduced into a conductor by narrowing its cross-sectional area, by poor splicing between two pieces of wire (see Fig. 1-3), or by bad mechanical joining of two conductors (e.g., loosely tightening a wire under a binding screw in making connection to a plug, switch, outlet, or junction box). The resistance introduced by each of these results in a drop in voltage or *voltage loss* as current attempts to flow. Such losses can introduce difficulties ranging from annoyance to hazard, depending upon how much current is flowing, because *every* resistance converts a portion of the current flowing through it into heat. Heat is electrical energy loss. And so, the amount of current that can flow through the conductor is limited and a drop in the available voltage results at the conductor end connected to a lamp, appliance, or other device.

Keep in mind the distinction between the devices such as lamps and heaters that purposely include resistance and the wire conductors that are merely intended to carry current and deliver voltage; wires should have the lowest possible

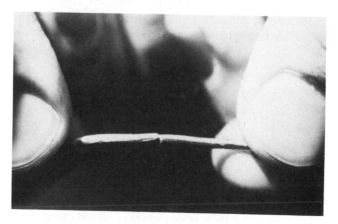

Fig. 1-3. A nicked wire has high resistance and therefore poor current carrying capacity. (Graf–Whalen photo.)

resistance, or they will waste energy. In addition, there is an obvious fire danger in having wiring heat up in the walls of a home. For this reason, home wiring is done with low-resistance, large-diameter, insulated copper conductors, as discussed in later sections.

The *ohm* (Ω), the measurement unit of resistance, is named in honor of the German physicist Georg Simon Ohm (1787–1854). In performing a lengthy series of experiments on the nature of current flow through different conductive materials, Ohm was the first to recognize resistance as an inherent property of matter. He later related resistance to voltage and current, demonstrating the mathematical balance between these three quantities in every electrical circuit. We talk more about the famous (and useful) law that bears his name after we discuss power.

POWER

Up to this point, we have seen that voltage comes from a difference in the quantity of electrons available between the two terminals of a source. Current is the flow of electrons through a conductor from the terminal having an abundance of electrons to the terminal having a scarcity of them. Resistance is the opposition offered to the flow of electrons by a conductor.

And now, let's look at *power*.

Power can be thought of as the energy equivalent of voltage and current in some other form. It is a measure of the work that results from their combined action in a resistance. Mathematically, power is the product of voltage and current. In direct current circuits, we simply multiply the voltage by the current and we obtain the power, expressed in *watts* (W).

The watt (named for James Watt, inventor of the steam engine) simply expressed the rate at which energy is being converted from one form into another, that is, *how we are putting voltage and current to work.*

Everybody uses the term "watts." You buy a light bulb by its wattage rating. And, you know that a 100-watt bulb gives you much more light (and gets quite a bit hotter) than a 25- or 50-watt bulb. Now, if all household light bulbs are intended to work from a 115-volt source, what's the difference between them? The answer is the resistance of their filaments (see Fig. 1–4). The 25-watt bulb filament has a relatively high resistance; it limits the amount of current that can flow and so the filament's atomic structure is subjected to less friction by the electrons flowing from the source. Therefore, it does not heat-up to the same intensity as the 100-watt bulb, which has a lower-resistance filament.

With four times as much current flowing through the 100-watt lamp filament, its atomic structure heats up to a much higher temperature. The result is simple; the light output of the 100-watt lamp is much greater, because *more work is being done;* its filament is converting voltage and current into *heat* (and, incidentally, light) at four times the rate of the 25-watt bulb!

Fig. 1-4. Both light bulbs shown operate from the same voltage though they have different wattage ratings. The bulb with lower resistance carries more current, gives more light and has a higher wattage rating. (Graf–Whalen photo.)

Power measurement in watts is thus a useful way of expressing how much work an electrically-powered device is doing in the real world. There are many uses for power measurement. Surely you have heard people talking about the power output of a motor. They speak in terms of *horsepower* (hp), and they are usually talking about the amount of mechanical output power it delivers. Now, to produce that mechanical power, the motor needs electrical power. And so, there is a direct relationship between the electrical power and the equivalent mechanical power. It is:

$$1 \text{ horsepower (hp)} = 746 \text{ watts (W)}$$

What does this mean in practice? Well, if you have a 115-volt motor that the manufacturer has rated at 1 hp, the source and conductors supplying electrical power to the motor must provide at least 6.5 amperes of current to achieve the rated mechanical output of the motor. Mathematically, that is:

$$1 \text{ hp} = 746 \text{ W} = 115 \text{ V} \times \text{amperes}$$

Solving for current:

$$\text{amperes} = \frac{746 \text{ W}}{115 \text{ V}}$$

$$\text{amperes} = 6.5 \text{A}$$

In practice, there are watt losses in motors that limit their energy conversion efficiency. This means that the electrical source must deliver more energy than the equation states. Usually, this is about 30% higher. The lost energy is converted into heat (which is why you may have noticed your electric drill getting warm as you use it). Those heat losses are inescapable inefficiencies in converting electrical power into mechanical power, because there's no such thing as a *perfect* power conversion device. Nevertheless, the relationship between electricity acting through a power conversion device and the mechanical work done is one of the useful measures that you will need to know in your home electrical work.

THE ELECTRICAL CIRCUIT

Earlier, we looked at the way a power source having some voltage between its terminals pushes a current through a conductive pathway connected between those terminals. All of these taken together, are called an *electrical circuit*. The name is an old one. Once, the word "circuit" was used to describe a race track at which horses would start from a point, race about a closed, loop-shaped pathway, and finish at the same point they started. Perhaps some early experimenter was struck by the similarity to the way current raced around a closed conductive loop, starting from a point and ending at the same place. Nevertheless, the term has endured and today we speak of a circuit being *closed* or complete when a conductive pathway exists from one terminal to the other terminal of a power source. The circuit is *open* or incomplete if the conductive pathway is interrupted by a nonconductor (such as air). An open circuit simply does not allow a current of electrons to flow. The most common way of interrupting a circuit is with a device called a *switch,* which is discussed later.

There Are Only Two Kinds of Circuits

When only one conductive pathway exists between the terminals of a power source, but that single path is through several connected devices, the circuit is known as a *series circuit*. Figure 1-5 illustrates a series circuit, made up of three lamps. As you can see, the only way that current can flow through this circuit is to pass through the filament of lamp 1, then that of lamp 2, then that of lamp 3, and back to the other terminal of the power source. The principal characteristics of a series circuit are:

1. The *same* current flows through each device in the circuit.

2. The voltage *decreases* across each device, in proportion to its resistance. (Let's say that all lamps are identical, two-volt types, and the voltage between the terminals of the source is 6 volts, measured at A. If we measure at B, the voltage

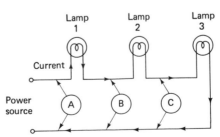

Fig. 1-5. Series circuit.

is 4 volts. If we measure at C, the voltage is 2 volts. Thus, the source voltage has decreased by just the amount required to get each series lamp to glow at full brightness.)

3. An *interruption* anywhere in the series circuit will stop current from flowing.

Next, we come to the *parallel circuit*. With a parallel circuit, two conductors extend the terminals of the power source to any number of devices connected between them. Figure 1-6 illustrates a parallel circuit made up of three lamps. In such a parallel circuit, there are three branches; that is, each lamp serves as a separate current path, independent of the other two. Across each lamp, we measure the full supply voltage. This means that we can now connect lamps that require different currents to work properly, as long as each lamp is rated at the voltage of the power source.

What total current flows from the source? Simply, the sum of the branch currents. With a parallel circuit, it does not matter if one branch takes 1 ampere, the second takes 5 amperes and the third takes 3 amperes. The total current flowing from the source will be 1 + 5 + 3, or 9 amperes. But, the currents will divide between the branches in proportion to their resistance.

The principal characteristics of a parallel circuit are:

1. Independent branch currents flow through each device in the circuit.

2. The voltage across each device is the same, although total current divides in inverse proportion to the resistance of the branches. (In this example, let's assume that all lamps are different. Each is rated at six volts, and that is the voltage of the source. But, lamp 1 requires one ampere, lamp 2 requires 5 A, and lamp 3 takes three amperes. All glow at rated brightness, because the parallel arrangement makes each one seem to be the only device powered by the source.)

3. The *total current* taken by all branches flows from the source, but divides between the branches.

4. Interrupting just one branch does not stop current flow through the others. To open this entire circuit, the *supply line* connecting all branches to the source must be opened.

Although the two types of circuits just described are pure cases of series and

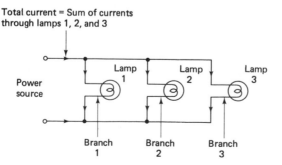

Fig. 1-6. Parallel circuit.

parallel arrangements, there are a great number of combinations that are referred to as *series-parallel circuits*. As we discuss later, your home wiring system comprises many branch circuits. These circuits always have a circuit disconnect device (such as a fuse) in series with a network of parallel circuit paths back to the source, making combination series-parallel circuits, as we discuss shortly.

Shortly? Oh yes. That brings up the topic of the so-called *short circuit*. It really isn't a kind of circuit, but a kind of problem. A *short* is a direct connection between the two terminals of the power source. It can be caused by an accidental direct wiring connection or some other slip-up (such as an accidental contact between the bared ends of wires connected to the source), which causes high current to flow.

A short circuit is a problem because the current that flows is probably limited *only* by the tiny resistance of the wires leading to the source. The connection can thus allow such a huge current to flow that the wire instantly heats to a glowing red, burning and melting the surrounding insulation. Carried to an extreme, it can heat surrounding materials to the point where they flash into flame. Obviously, a short circuit is a kind of circuit you hope not to create. But, fortunately, there are protective devices, such as fuses and circuit breakers, to protect against these dire consequences in your home wiring system. These are covered in Chapter 2.

MR. OHM'S WONDERFUL LAW

Of all the early electrical experimenters who worked to sort out our understanding of electricity, Georg Simon Ohm deserves a special place in history. He devised a simple law that expresses the relationship between voltage, current, resistance, and power in an electrical circuit.

Ohm's law states that:

> *The strength of an electric current in a circuit is equal to the voltage divided by the resistance.*

Mathematically, that is:

$$\text{Current (amperes)} = \frac{\text{Voltage (volts)}}{\text{Resistance (ohms)}}$$

Using quantitative symbols, it is:

$$I = \frac{E}{R}$$

Why is Ohm's law so great? First, it lets us attach numbers to the ideas we have discussed previously. Second, it establishes resistance as a numerically definable quantity, using the ohm as its unit.

To see what this means, let's work out an example:

We have a circuit consisting of a 20-ohm resistance, connected to a voltage of 115 volts. What current is flowing?

$$I = \frac{E}{R}$$

$$= \frac{115 \text{ V}}{20 \text{ } \Omega}$$

$$= 5.75 \text{ A}$$

Without very much juggling, it is possible to use any two known quantities to arrive at the third, unknown quantity. Thus:

Voltage (volts) = Current (amperes) $\times$ Resistance (ohms)

In symbols:

$$E = I \times R$$

and

Resistance (ohms) = $\dfrac{\text{Voltage (volts)}}{\text{Current (amperes)}}$

In symbols:

$$R = \frac{E}{I}$$

By using Ohm's Law, we can clearly see how voltage, current, and resistance interact in an electrical circuit. We can predict that by increasing voltage, we will increase current flow through a given resistance or, by decreasing resistance, we will get a larger current from a given voltage source. The important feature is that Ohm's Law specifically tells you exactly how much of a change you will see, and that is what makes it truly wonderful!

Going a step further, the fact that Ohm's Law puts a recognizable value on resistance gives us another handy calculating tool to help us to arrive at the power expended in a circuit. As discussed earlier, power (watts) is the product of voltage (volts) and current (amperes). In symbols, that is:

$$P = E \times I$$

But, remember that voltage has a mathematical equivalent in Ohm's Law, which is:

$$E = I \times R$$

And so, we can rewrite the power equation like this:

$$P = (I \times R) \times I$$

OHM'S LAW: $E = I \times R$

In any given situation you may know any two of the quantities and want to solve for the third. A simple way to remember and use Ohm's Law is given by the chart below.

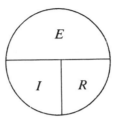

To find a desired item, cover it. The remaining two items shown are to be multiplied or divided to find what you are seeking.

For example, knowing I (current) and R (resistance), find E (voltage) by covering E with your finger. The chart shows that $E =$ $I \times R$ or $I \times R$.

To find I, knowing E and R, cover I. The chart shows that $I = E/R$ or $E \div R$.

or like this:

$$P = I^2 \times R$$

Interesting? Sure. Now it becomes possible to calculate power just by knowing the current, squaring it (multiplying it by itself), and multiplying by the resistance. You don't even have to know the voltage.

Well, if you can do it one way, can you do it another? Yes thanks to Ohm's Law.

Current also has a mathematical equivalent in Ohm's Law, which is:

$$I = \frac{E}{R}$$

And so, we can rewrite the power equation a second time, like this:

$$P = E \times \frac{E}{R}$$

or like this:

$$P = \frac{E^2}{R}$$

As in the first case, things have gotten easier. Now, power can be calculated just by knowing the voltage, squaring it (multiplying it by itself), and dividing by the resistance. With this approach, you don't need to know the current.

ELECTROMAGNETISM

Everyone is familiar with a magnet. It is a fascinating piece of metal surrounded by an invisible "force field" that exerts a pull on any nearby metallic object that is made of iron or steel. The most common shapes for magnets are the *bar magnet* and the *horseshoe magnet* (which is really a bar magnet with its ends bent close to one another). The field surrounding a magnet is most concentrated at its ends. These are labeled *N* (north) and *S* (south) poles, by scientific convention (see Fig. 1-7).

Perhaps you have noticed that when two magnets are brought together, they will *repel* each other if held one way or exert a *strong pull* on one another if the direction of one is reversed. That is, if two N poles or two S poles are brought together, repulsion occurs. Yet, if an N and an S are brought together, a strong attraction results. Here again is another manifestation of the law we learned earlier: *Like charges (or poles) repel; unlike charges (or poles) attract.*

What has this to do with electricity? Well, one of the fascinating properties of electricity is its ability to establish an invisible field of magnetic force about a conductor. Just as a magnetic field surrounds a bar or horseshoe magnet's poles, so a conductor carrying current is surrounded by a magnetic field that is perpendicular to the conductor and detectable throughout its entire length.

The remarkable difference between the physical magnet and the electromagnetism resulting from current flow through a conductor is that the magnetic field surrounding a conductor becomes *stronger* if the current flowing through the conductor is increased. (In fact, the strength of the electromagnetic field is directly proportional to the amount of current flowing through a conductor at any

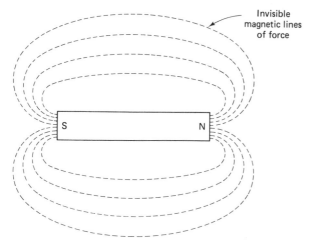

Invisible magnetic lines of force

Fig. 1-7. A bar magnet is surrounded by invisible lines of force that circle through space from the N to S pole. Electrons flowing in a conductor produce similiar force fields around the conductor.

instant.) Because of this property, it is possible to convert electric current flow into mechanical energy using the magnetic attraction/repulsion of the electromagnetic field surrounding a conductor. This is the basic principle incorporated in electric motors, solenoids, relays, and other electrically-powered motive devices.

There is an even more fascinating link between magnetism and current flow in a conductor and that is the principle of *electromagnetic induction.* Sounds complicated, but it's not. Suppose we take a length of insulated wire and wrap it around a hollow cardboard tube so that we form a single layer coil of several dozen turns. Next, we connect the two free ends of the coil to a sensitive current-measuring meter (scientists call it a *galvanometer*). Then, we move a bar magnet in and out of the coil's hollow center. The needle on the meter swings back and forth. But, there is no battery or other current source in the circuit, so why is the meter saying a current is flowing? Simple. As the bar magnet is moved into and out of the coil, its magnetic field is passing through the conductor's turns. The moving field is exerting a force on the electrons within the conductor's atomic structure, causing some to move. When electrons move, a current flow is produced. And so, what we have here is a simple *generator.*

Let's try another experiment. Suppose we formed a square loop of wire and placed it between the poles of a strong, stationary horseshoe magnet. To the free ends of the loop, we attach our meter. Now, we rotate the loop, and . . . whoops! There it is again . . . current! (See Fig. 1-8.)

Two things are obvious from these simple experiments: It doesn't matter whether it is the magnet field or the conductor that is moving. The important fact is that a change in the intensity of a magnetic field as it passes through a conductor will induce a current flow in the conductor. In addition, the more rapidly the change occurs, the greater the current flow!

Are you ready to take these thoughts a step further? Okay. Let's take a battery and connect it in series with a switch, joining their free terminals to the ends of a coil of many turns of wire, which we will call coil A. Now, on top of coil A, let's wind another coil of insulated wire, which we will call coil B. The free ends of coil B connect to our sensitive meter.

Now, open and close the switch fairly rapidly . . . and the meter needle swings back and forth, indicating that a current flow is taking place in coil B! How can this be? Well, each time we close the switch, battery current flows through coil A, building up an electromagnetic field about that coil but also embracing the turns of coil B. When we open the switch, current flow ceases and there is nothing to sustain the field, so it immediately collapses. As it collapses, it passes through the turns of coil B, inducing a brief flow of current. It doesn't matter that the coils are electrically insulated from each other; they are electromagnetically coupled so that a change in current flow in coil A brings about a responsive current flow in coil B.

You might think of this as an *energy storage* situation. Current flowing into coil A results in energy being stored in the magnetic field about the coil. So long

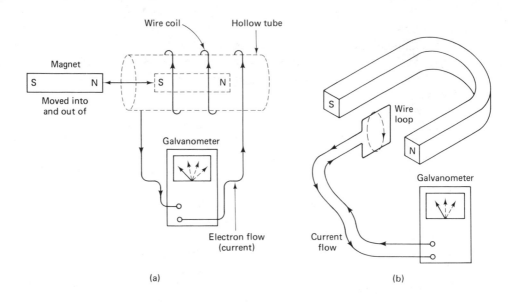

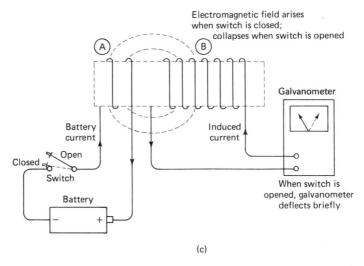

Fig. 1-8(a). Moving a magnet into the coil induces a current flow in the coil, causing the galvanometer pointer to deflect. **(b).** If you twirl a wire loop between the poles of a horseshoe magnet, the galvanometer also deflects, indicating current flow in the wire. **(c).** How current is induced from one coil to another, purely by the electromagnetic field.

as that current doesn't change (as it wouldn't, if we are using dc from a battery), nothing happens in coil B. But, when we disconnect the current source by opening the switch, the energy stored in the field has to be returned to the circuit. So, as it collapses, the field transforms that energy into a current in coil B.

Two other names for the crude two-coil device we have used in this example are: an *induction coil* (which is a d-c device, rather like the spark coil that energizes your car's ignition system); and a *transformer* (which is an indispensable part of the massive a-c power distribution system that serves our homes, offices, shops, and factories).

Two subjects that have not been discussed here are *how much current* or *how much voltage* are induced into coil B by the magnetic field about coil A. The answer depends upon several factors, but the most important one is the ratio of the number of turns of wire that make up coil A to the number that make up coil B (see Fig. 1-9). In this, we find the secret to the idea of *stepping up* or *stepping down* in an a-c system. (These casually used terms can be confusing. But, let's dispel the clouds of mystery.)

If coil A and coil B have the *same* number of turns, their turns ratio is 1:1. This means that if a voltage of 1 volt is applied to coil A and some current flows through it, the voltage induced across coil B will be 1 volt, at a corresponding current. But, let's say that we want a *higher voltage,* say, 10 times the voltage, to appear on coil B. We must wind 10 times as many turns of wire on coil B. The ratio of A to B is now 1:10. And so, with 1 volt applied to A, we get *10 volts* across B! Sounds great, doesn't it? But, here's the kicker: The power product of voltage and current into coil A *has to equal* the power product of coil B. So, if we get 10 times the voltage of A in B, we only get 1/10 the current! Conversely, if we want 1/10 the voltage of A, we can get 10 times the current to flow in B.

If all of this sounds like an electrical proof that there is no such thing as a free lunch—you're right! Transformers and turns ratio juggling are just devices by which we can rearrange the relationships between voltage and current in two magnetically coupled circuits. But, there is no way that a transformation device can *create* power. So, when we speak of "stepping-up" or "stepping-down," you

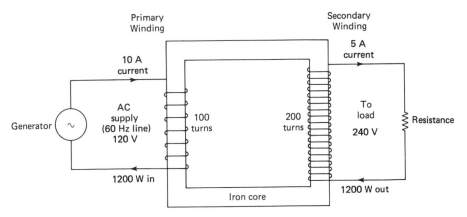

Fig. 1-9. How a transformer works. The continuously changing current through the primary winding results in a continuously changing electromagnetic field. The field's force is coupled to the secondary winding through the iron core. There is little energy loss. The electromagnetic force induces a voltage in the greater number of turns of the secondary, and current flows to the load. Transformers make it possible to transmit power over long lines very efficiently and to step-up or step-down voltage with ease.

should read it as a manipulation of voltage and current to achieve some result. There is no way that a transformation device can create energy!

THERE ARE TWO KINDS OF CURRENT

To simplify our discussion, we have talked about current flow as if it was always in one direction, that is, from the power source terminal at which there is a surplus of electrons to the terminal at which there is a scarcity. In terms of time, however, this explanation is only true for a battery or other source that produces the one-way electron flow we call direct current (or dc).

Direct current flows in just one direction, from the negative terminal of the source, through a connected circuit, to the positive terminal. Why negative to positive? Well, because all electrons have a negative charge. Where there are few or no electrons, that location just has to be positive with respect to the location where there are many. Have you thought just the opposite? That is a natural but erroneous conclusion. You are in good company, though. Two centuries ago, Ben Franklin made the same mistake and started people thinking that current flowed from positive to negative. This is not true, however, as science has since proven. But the quaint belief that Franklin started was given the title *conventional current flow* to distinguish it from *electron current flow,* which is the real way that current flows. Does it really matter? Not greatly. But, visualizing a flow of negatively-charged electrons moving from the most negative (−) to the less negative (+) terminal of a source seems simpler, and it is more correct.

A d-c source has a certain voltage that may be produced by a chemical reaction, as in a battery, or by conversion of rotational mechanical energy, as in a d-c generator. Either way, the voltage always ends with one terminal negative, the other positive. When a circuit is connected between these terminals, the current that flows is dependent only on the resistance of the circuit. The greater the resistance becomes, the smaller the current.

Back in 1879, when Thomas A. Edison invented the electric lamp, dc was the only kind of current man could make. Edison envisioned a great scheme to replace gas and kerosene lighting with his new lamps, which would be powered by conveying d-c electricity through wires. It was only after he had wired a small area of Lower Manhattan in New York and started his d-c generators at the famed Pearl Street Station that Edison began to see some of the shortcomings of dc. Here was the problem: Edison created a miles-long parallel circuit, with branch circuits in each building along the way. The problem was that the main wires leading to the d-c generators, even though of pure copper and of large diameter, had enough resistance to limit the current available to customers at the end of the circuit.

What a headache! If Edison adjusted his d-c generators so that customers nearest his powerhouse received the right voltage at their lamps, the resistance in the main wires would reduce the voltage delivered to customers at the end of the circuit so that their lamps glowed dimly. If he raised the voltage to satisfy the customers at the end of the line, those nearest the powerhouse got too high a voltage, their lamps burned brightly but briefly, and more complaints poured in. Eventually, Edison compromised his way out of this problem, but it helped to show that d-c power distribution had serious shortcomings.

It was not long after that the brilliant idea came up of a power source that periodically reversed its polarity. Such a source would produce alternating current (ac); that is, a current that flowed in one direction for a fraction of a second, then reversed polarity so that current flowed in the opposite direction for another fraction of a second to complete a *cycle.*

Why was this such a brilliant idea? Because ac can be centrally generated at a power house, stepped-up to a high voltage, transmitted over long power lines, then, stepped-down at each location where power is needed, all with vastly less loss of power than would be possible with dc. It is this collection of virtues that makes possible the enormous power distribution networks throughout the world. And, it is ac that is supplied to your home.

Today's a-c power is precisely generated by huge machines known as *alternators.* These are special a-c generator that convert energy supplied by falling water, burning coal, oil, natural gas, or the heat of atomic reaction into mechanical, rotational energy. This, in turn, is converted into equivalent electrical power. in which the polarity of the voltage (and, hence, the direction of current flow) completes a full polarity reversal (cycle) every 1/60 of a second. Thus, the standard North American a-c power runs at 60 cycles.

A CLOSER LOOK AT ALTERNATING CURRENT

There is nothing mysterious about ac. It is just another way of producing electric power. But, it is different enough from dc to warrant a more detailed discussion.

There are instruments, called *oscilloscopes*, that let us examine the shape of electrical waves in relation to time. If you were to use such an instrument to examine the a-c power line, you would see on its screen a graphic tracing that looks like Figure 1-10. This is a trace of a-c voltage. The shape of the tracing is known as a *sine wave.* That is because it follows the mathematical characteristics of a sine function.

Commencing from a mid-line zero voltage point, the a-c voltage wave rises smoothly to a positive value, then falls just as smoothly back through zero and reaches a negative value, smoothly rising again through zero to a positive value, and so on. This waveform clearly shows that a-c voltage reaches a peak positive value, as well as a peak negative value, in one cycle. During the time that it is rising toward its positive value, electrons will flow in one direction through a conductor connected to the source. But, as the voltage wave "turns around," electrons will flow in the opposite direction, until the wave reverses again.

So what have we accomplished? Well, we have made a current flow through some connected device, first one way, then the other. What result has been achieved? If the device is a lamp, its filament has become just as hot as if current were flowing in only one direction, as with dc. If the device is a *transformer*, we have made it possible to step the voltage (and current) up or down, simply be-

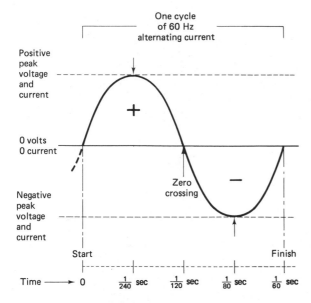

One cycle
of 60 Hz
alternating current

Positive
peak
voltage
and
current

0 volts
0 current

+

Zero
crossing

–

Negative
peak
voltage
and
current

Start

Finish

Time ⟶ 0 $\frac{1}{240}$ sec $\frac{1}{120}$ sec $\frac{1}{80}$ sec $\frac{1}{60}$ sec

Fig. 1-10. One cycle of a 60Hz AC power line voltage waveform.

cause we have changed the current flow by periodically reversing the polarity of the source.

And so, what we have achieved by supplying ac rather than dc to the home is a more versatile power source. We are able to make use of the changing electromagnetic forces that occur when ac flows that dc simply does not have. In that difference lies the key to the success of the enormous power distribution networks that serve our lives.

How A-C Power Reaches Your Home

The a-c power that serves your household needs is generated continuously, day and night, by the huge alternators at the power house. These great machines produce ac at a voltage of 13,800 volts, alternating at 60 cycles (or, as it is now more proper to say, 60 *Hertz*, in honor of the early electrical experimenter, Heinrich Hertz. It is usual to abbreviate this word to *Hz*, although the meaning remains *cycles per second*.)

Why do power alternators operate at such a high voltage? Because it is current, not voltage, that causes power to be wasted in transmitting electrical energy through conductors. If a high voltage is used, current can be reduced through transmission lines, thus reducing energy loss.

Note: *Power is the product of voltage multiplied by current.*

A voltage of 13,800 volts at a current of 1 ampere yields a power capacity of 13,800 watts. If we were using a lower voltage, say 115 volts, it would mean a large current flow of 118 amperes is required to achieve the same power. Because power loss increases as the square of current through a conductor, if the conductor has a resistance of 1 ohm, the power loss in the high-voltage–low-current circuit is only

$$
\begin{aligned}
\text{Power} &= \text{Current}^2 \times \text{Resistance} \\
P &= I^2 \times R \\
&= (1\ \text{A})^2 \times 1\ \Omega \\
&= 1\ \text{W}
\end{aligned}
$$

But, in a low-voltage (115-V) circuit a current flow of 118 amperes through the same 1 ohm conductor resistance would produce a power loss of

$$
\begin{aligned}
\text{Power} &= \text{Current}^2 \times \text{Resistance} \\
P &= I^2 \times R \\
&= (118\ \text{A})^2 \times 1\ \Omega \\
&= 13{,}924\ \text{W}
\end{aligned}
$$

Keeping current to a minimum means less loss of energy in transmission, which is why power company alternators work at such a high voltage (and, in some cases, even higher).

The a-c power produced by the alternators is next stepped-up to even higher voltages (ranging from 23,000 to 765,000 volts) through large transformers outside the power house, depending upon how much power must be transmitted over how great a distance. In a city, where transmission distances are relatively short, the voltage step-up is small. But where power must be transmitted over hundreds of miles of conductors, the maximum voltage step-up is used to keep current very low. Figure 1-11 shows a simple view of a distribution system of this type. In some urban areas, residential distribution systems operate at 13,800 volts and above.

Wherever power is to be delivered from the transmission line, there is a substation with transformers that step-down the voltage to some lower value. For distribution purposes, this voltage is usually 2,300 to 4,000 volts, which is the voltage on lines that serve local areas (e.g., overhead lines at the top of utility company poles outside your home).

In the immediate area of local residential power users, there is yet another transformer, usually mounted on the utility company pole (see Fig. 1-12). This transformer steps down the 2,300- to 4,000-volt local transmission voltage to 115/230-volt residential voltage.

As you can see, the voltage has been stepped-down along the way to the final value needed by the residential customer. Only when the point of delivery has been reached has the final step-down been accomplished, so that relatively high current needed by local appliances flows only through the fairly short length of wire that connects the utility company's transformer to the in-house power distribution center of the home-wiring system.

The lines that serve your home can be either the *two-wire, 115-volt system,* common in small homes built before 1940, or the more modern *three-wire, 230-*

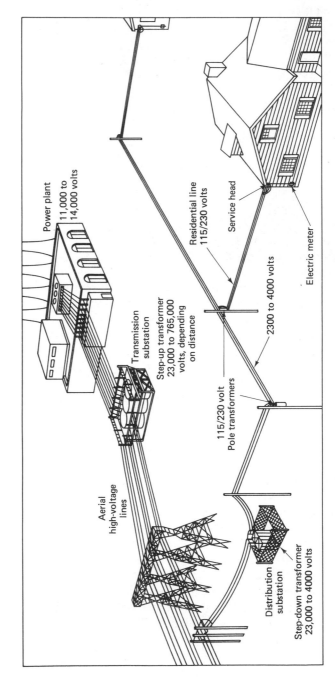

Fig. 1-11. From utility company power plant to the service head of your home. AC power undergoes step-up and step-down in voltage.

Power plant
11,000 to 14,000 volts

Transmission substation

Step-up transformer 23,000 to 765,000 volts, depending on distance

Aerial high-voltage lines

Distribution substation

Step-down transformer 23,000 to 4000 volts

115/230 volt Pole transformers

2300 to 4000 volts

Residential line 115/230 volts

Service head

Electric meter

Fig. 1-12. A pole transformer. (Graf–Whalen photo.)

volt system, which has become the universal choice in newer homes due to the increased use of appliances, such as air conditioners, dishwashers, dryers, and so on.

Two-wire system. Two-wire systems were meant to serve only lighting requirements of the home and so, the amount of power they can deliver is limited. Most have a maximum current rating of 30 amperes, which, at a voltage of 115 volts, gives the house a total power of 3,450 watts. This is not high by present standards.

Commencing from a two-wire connection to the service head of the house, wires run to the utility company meter, then to the fuse box. The two wires are color-coded black and white. The black wire (also called *hot* or *live*) is completely insulated. But, the white wire (also called *neutral* or *ground*) is connected to a water pipe or some other metallic object that is electrically in contact with the earth. Current flows equally through both wires. The reason for *grounding* the white wire is to place all the parts in your home electrical system at the same potential (that of the ground you stand on or contact in everyday use). The system is thus self-protected should a problem, such as excessive voltage, develop in the utility company's power circuit.

The two-wire system is antiquated by present standards and imposes limits on the enjoyment of modern labor-saving appliances because it simply does not make enough current available to the home. Changeover to the more modern three-wire system is an easy matter that will update the two-wire home to livable standards in today's electrical world. It is an expense but also an investment, both in better living and in resale value for the older home.

Three-wire systems. Most homes today are equipped with the three-wire system because it provides both 115– and 230-volt power to the home. This allows a much broader choice of appliances. Concern for adding new circuits or devices is lessened simply because there is more power capacity with which to work.

You can identify a three-wire system by counting the wires connected to the service head by your utility company. The three wires consist of a neutral, white wire and two hot or live wires, one being black, the other red. The voltage between the neutral wire and each of the hot wires is 115 volts. But, the voltage between the black and red wires is 230 volts. The size of all the lead-in conductors in new home construction is usually great enough so that, typically, 100 amperes at 230 volts can be supplied through the utility company's connecting wire, a total of 23,000 watts. The best guide is the rating of the main circuit breaker. If it is less than 100 amperes, you have a three-wire system with only moderate current-carrying capability in the lead-in wires. These would have to be changed to larger conductors if you intend to upgrade to more modern standards.

2

SAFETY

Electrical Codes–Grounds–Ground Fault Protection

Each year in the United States, a thousand or more people are injured or die as a result of careless use of electricity or as a consequence of using faulty or improperly installed electrical equipment. In addition to personal hazards, there are millions of dollars of property damage. When working with electrical equipment and installing systems for your family's use, you cannot overlook these statistics. Electricity is a form of energy that can be tremendously destructive or lethal if treated with less than reasonable care and utmost respect. But, as we learned in Chapter 1, electricity follows predictable paths and a logic that makes it the most manageable of all energy sources. Local governments have adopted formal codes whose objective is the enforcement of proper installation to ensure the safety of you, your family, your neighbors and your community. By understanding the logic behind the safety rules, you will see why it is necessary that the codes be followed to the letter.

ELECTRIC SHOCK

As you read this, your body is performing countless functions, such as blood circulation, respiration, digestion, holding this book, and controlling the eye muscles, to name a few. In all of these activities, there is an internal electrical activity associated with each contraction of every muscle involved. Yes, the human body relies on tiny, self-generated electric currents for its operation. Muscles, nerves, the brain, and major organs all operate from their own "batteries" of sodium and potassium ions.

It is precisely because the human body operates electrically that it is susceptible to injury or destruction if an external source of electricity causes a significant current to flow through the body. Such a phenomenon is known as *electric shock*. Shock can range from an unpleasant tingling sensation at the body surfaces in contact with the source to unconsciousness and paralysis of the muscle masses that control breathing or even a disoriented twitching of the heart muscle that could prevent respiration and circulation. The degree of bodily reaction to an electric shock depends on the amount of current absorbed. The greater the current, the worse its effects on bodily functions and the greater the risk of death. A normally healthy adult can be fatally injured by as little as sixty-thousandths of an ampere [60 milliamperes (mA)] of alternating current passing through his chest (adults in poor health or children may be killed by even less). This would probably occur in the so called worst-case situation in which electricty enters one hand and leaves via the other. Current therefore flows right across the chest and has the most potential for affecting the heart and respiration (see Fig. 2-1).

Several factors determine the severity of an electric shock: the resistance of the body (lowest when it is wet); the voltage involved, which will determine how much current passes; the path of the current; and, how long the current flows. As just mentioned, currents are particularly dangerous when passing through the chest from one arm to the other, or even from one arm to a foot, especially if your body is partially submerged in water (in a tub, pool, or the like). Current entering one finger and leaving by another on the same hand, however painful, is generally less dangerous. Remember though, a shock of this sort can be unpleasant enough to cause other problems, such as a fall from a ladder caused by your reaction. In other words, *no amount of current is too small to ignore*.

A human body that becomes part of an electrical circuit conforms directly to

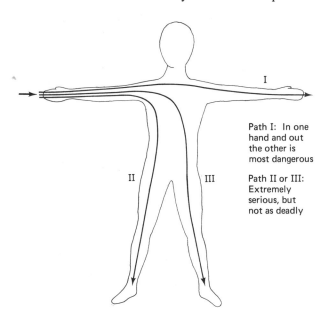

I

Path I: In one hand and out the other is most dangerous

Path II or III: Extremely serious, but not as deadly

II   III

Fig. 2-1. The path that electric shock current travels can make the difference between life and death. Current flow through the chest can disrupt heart function and paralyze breathing muscles. Keep yourself out of the circuit!

Ohm's law: Current is high if the body's electrical resistance is low and low if the body's electrical resistance is high. Minimum body resistance between major extremities is about 500 ohms. However, surface or contact-resistance at the skin, the chief initial current-limiting factor, varies widely. Dry skin resistance can be high (over 100,000 ohms). Wet skin may be only 100 ohms. Assuming the source is a 120-volt house circuit, accidental current flow through the body from hand to hand can easily vary from less than two milliamperes for dry hands (barely enough to be unpleasant) to a life-threatening 150 milliamperes or more if your hands are wet or sweaty!

Another factor in shock is its duration. You might have heard of people grasping a live wire and being unable to release it. This is the most dangerous situation of all. This does not mean that a live wire will stick to you, as if by magnetism. Rather, when your hand touches a live wire and surrounds it, even a low level current, perhaps as low as 1/10,000 of an ampere, the current will cause your muscles to contract and the hand to clench so tightly that you cannot open your fingers to release the wire. With this situation, if current flows for many seconds, the shock can easily become fatal by stopping heart action or breathing. At very least, severe burns will be inflicted, which are serious enough to warrant emergency attention.

NATIONAL ELECTRICAL CODE*

The first nationally recommended wiring code was published in 1895 by the National Board of Fire Underwriters. It became the basis of the National Electrical Code (NEC) that was drawn up two years later through the efforts of insurance, architectural, electrical, and related groups. Today, expanded and amended many times to keep pace with the ever-greater use of electricity, the NEC is the standard of the National Board of Fire Underwriters and is recommended by the National Fire Protection Association (Fig. 2-2).

The Code aims to enforce use of safe wiring materials, devices, and practices by establishing minimum standards to ensure standardization and proper installation. It is not a wiring instruction manual, but it specifies the correct materials to use under various conditions and it defines the correct methods of using them. The NEC does not have, in itself, the force of law. Adoption and enforcement is left to the discretion of local authorities. When a community adopts the NEC as a part of its local regulations, it acquires legal force. Often, too, the NEC is combined with a local code that may be even more strict. Also, local codes vary. For example, one city may bar homeowners from doing any wiring work. Another may not only permit it, but offer encouragement and advice. Yet, another may impose a considerable inspection and permit fee, and still another municipality may have no regulations whatever. Check your local

*You can purchase a copy from the National Fire Protection Association, 470 Atlantic Avenue, Boston, Massachusetts 02210. You can order the abridged version, called "One and Two-Family Residential Occupancy Electrical Code" from the same source.

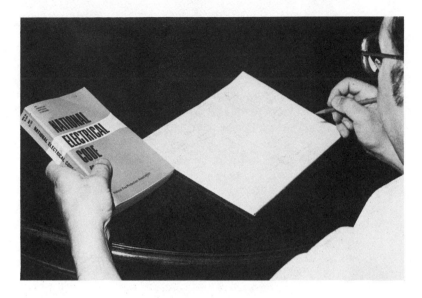

Fig. 2-2. Keep your current NEC handy for referral when you plan to improve or extend your home's existing wiring. (Graf-Whalen photo.)

electrical code, because any work you do must meet its requirements *to the letter*.

In general, it's a good idea to have a current copy of the NEC on hand if you plan any wiring work. And, of course, a copy of the local code should be with it as well. You will probably find that a permit is required before starting any electrical work and that an official inspection must be completed before your work is approved, or indeed, before your local utility will agree to supply power. Also, even if you get away with it, if you had a fire of electrical origin, your insurance company would be within its rights to refuse payment for a fire caused by noninspected work. So, at all times, play it safe and follow the rules set down by your local authorities.

Don't let the thickness of the printed NEC book mislead you into feeling that looking up a particular item will be complicated. A considerable portion of the book is devoted to special problems (and their solutions) likely to be encountered in industrial and commercial work and other nonresidential wiring. It is worthwhile for the prospective homeowner-electrician to read through the Code's rules for residential work. The language is plain and clear-cut, and the recommendations are logical.

Note carefully that all wiring techniques and design criteria specified in this book are interpretations of the latest NEC requirements. You must recognize that, although every effort has been made to interpret correctly, there can be no acceptance of liability for improper installation because the codes and your local inspector are the final authority.

Use Only Approved Materials

Besides checking the codes, national and local, for proper installation methods, you must also determine whether there are restrictions on the purchased materials that you may use. Many local codes say simply that materials must be found acceptable by a "nationally recognized testing laboratory," others go further and specify UL or Underwriters Laboratory listing only (see Fig. 2-3). What does this mean?

Like the National Fire Protection Association that produces the NEC, Underwriters Laboratory is a nonprofit, nongovernment, industry-supported organization. It has no legal enforcement power. However, it does test electrical products submitted by manufacturers for foreseeable safety hazards to both people and property. If a product meets UL standards, it is "listed," and the manufacturer is given permission to mark each production unit of the product with the UL label.

Contrary to popular belief, UL listing does not mean that the product is safe for any purpose people might think up. It does mean that the product is safe when used for the purpose for which it was intended. An inspector may therefore disapprove a listed product if it is used improperly; for example, the NEC does not permit the use of armored cable in barns. Also, listing does not mean that different components or devices are of the same quality. It merely means that both meet the *minimum* UL safety requirements. Durability, ease of installation, or extra features can make one a better buy than another for the same job. Your judgment must come into play, just as in buying any other product. In summary,

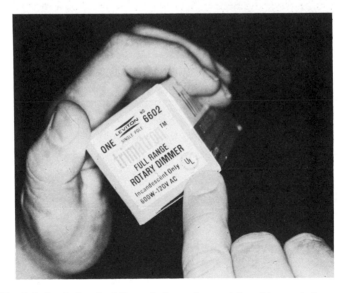

Fig. 2-3. Look for the UL symbol on all materials, wiring, switches or other devices that you purchase for your system. (Graf–Whalen photo.)

though, even if your local code does not require UL products, or if there is no formal code, do yourself a favor and stick to UL-listed electrical goods wherever possible.

GROUNDS AND SYSTEM GROUNDING

Now, hopefully, you are convinced of the need for safety consciousness and have determined to study the requirements of the applicable codes. The first and most important point to learn about the actual system is the procedure called *system grounding*. This critical safety practice involves two aspects: first, grounding at the service entrance of the system neutral wire, literally, connecting the neutral wire directly to the earth outside your house foundation; and, second, grounding (bonding) all nonfunctional parts of the system branches back to the entrance ground. Most cables for 115-volt service contain a bare or green covered ground conductor as a third wire that is expressly designed for this purpose. Note that nonfunctioning parts of the system means outlet boxes, switch boxes and plates, fixtures, and other housings inside which electrical conductors are spliced or connected to working elements. Because connections are the most vulnerable system points, these housings need special care.

Let's examine this piece by piece. The NEC and all local codes based on it require that the system neutral wire be grounded. Once that is done, all parts of the house from roof to basement floor are electrically connected to the neutral wire, to a greater or lesser extent depending on the resistance between that part and ground. The service entrance ground must be made by connecting a conductor from the neutral bus bar in the distribution box directly to a metal cold water input pipe and to a supplementary rod or rods driven into the ground outside your house (see Figs. 2-4 and 2-5). The water pipe, from well or municipal supply, provides a good contact with the earth (ground) and makes a handy choice. Remember also that metal pipes are good conductors and if the entrance is grounded to the water supply there is a good ground connection to all parts of the house plumbing; this is the reason why bathtubs are particularly dangerous in terms of possible electric shock. When a cold water supply pipe is not available, the ground connection may be made to the rod(s) driven into the ground outside the house.

Why is this done? Well, imagine what would happen if your system was not grounded. A high voltage line accidently falling across the wires into your house could give a lethal shock to someone inside; all appliances connected to the system could be ruined as well. Also, lightning striking at or near the power company's pole could ruin your wiring and appliances by induced currents. Or, if the pole transformer fails, your system could suddenly receive full transmission potential of 1300 to 2500 volts. Proper grounding of your system allows all these dangerous potentials to bleed off safely into the earth. Bonding all electrical device enclosures to the "system" ground surrounds the electrical connections with a conductive surface that is at the same neutral potential as the earth you stand on. In this way, touching a metal switchplate or an appliance case and a

Fig. 2-4. Connect a jumper from the service entrance neutral to the cold water input pipe as shown above. (Graf-Whalen photo.)

Fig. 2-5. Even if you ground your service entrance neutral to a cold water input pipe, the connection at the water pipe must be continued by a bonding jumper to a ground rod driven into the earth outside the house. (Graf-Whalen photo.)

ground connection, such as a faucet or bathtub, does not cause a shock hazard.

Grounding the boxes, switchplates, and fixtures provides a protective shield around the wiring connections within. A short circuit from a faulty hot line to the metal box will allow a large current to flow back through the ground connection. This current will trip the circuit breaker or blow the fuse in the hot line to open the circuit. Until the overcurrent protector acts, all current flow will be contained within the metal terminal box or grounded metal equipment housing, and fire hazards will be greatly reduced. (See Fig. 2-6.)

To explain this more completely, let's use a picture. Figure 2-7 is a schematic presentation of how electricity works for you. The load can be a lamp plugged into a wall outlet, a ceiling fixture with several lights, a vacuum cleaner or blender motor, or any other electrical device you can think of. The load can also be many individual loads on a single circuit. We have just shown one and left out switches and other controls for simplicity. At the service entrance in your house, 115-volts is applied across two wires so that current can flow to the load and return in a

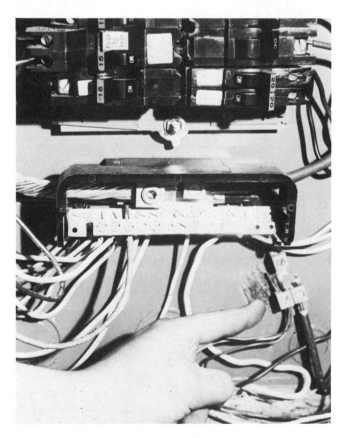

Fig. 2-6. The ground bus of the service entrance panel. All ground and neutral wires are tied together electrically. **(Graf–Whalen photo.)**

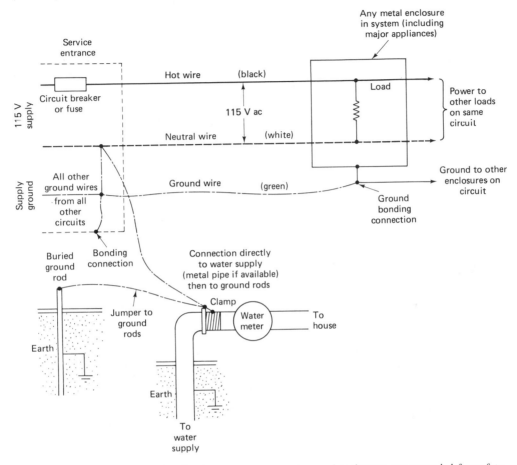

Fig. 2-7. Simplified diagram showing how your power system and enclosures are grounded for safety.

complete circuit to accomplish work. As we have said, the NEC requires that one line must be grounded (most likely to the house water supply as shown). When this is done, the grounded wire, called the neutral wire, is always at zero potential.

What happens now in operation? Whatever the load may be, we want the electricity to only flow to it, through it, and back with no detours or side paths. Ideally, electricity should never flow anywhere except in the black hot and white neutral wires and the load. Forgetting the green ground wire or bonding connection for a moment, consider what would happen if you accidently touched the wires. If you touch the bare end of the neutral there is little or no danger. You are standing on the ground and the neutral wire is at ground potential. There is no voltage difference so no current flows through you. Touching the bare end of the black hot wire is a different matter. It is at 115 volts above ground potential. If you touch it, the current has only one path to ground and you are it! Current will

33

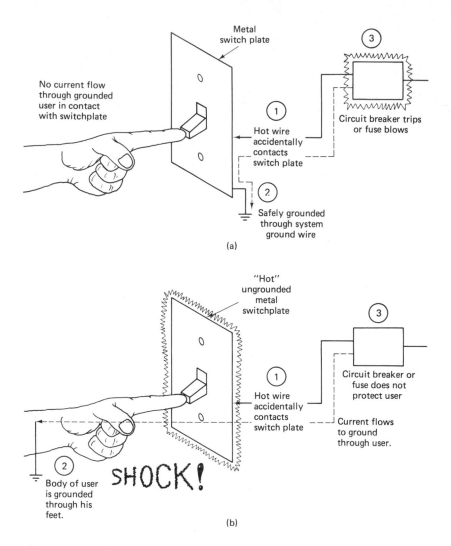

Fig. 2-8. Grounding protects you in case the hot wire contacts something you can touch. Without it, a lethal shock can be delivered with no protection from fuses or circuit breakers. (Remember: you can be injured by currents way below those that will blow a fuse or trip a breaker!)

pass through you. (See Fig. 2-8a.) The severity of the shock you will receive depends on how low your body's resistance to ground is, in other words, how much resistance there is between the hot wire and the ground connection. If your body is wet or you are standing on a damp, concrete basement floor, or worse, if you are in a bath and the ground is by way of the house plumbing, you represent a very low resistance and will be in serious danger if you come in contact with the hot wire. One part of your installation job will be to minimize these possibilities.

Luckily, it is not likely that you will ever touch the hot wire because it is normally confined in insulated cables and connections are safely made in outlet and junction boxes. However, it is possible that insulation can be defective or that a connection can come loose and allow a hot wire to touch an outlet box. What happens then? Figure 2-8b tells the story. If a defect of some sort allows current to flow to a surface (outlet box, switch plate, etc.) that you can touch, there could be a shock or fire hazard. Fortunately, the ground conductor provides a *second* current return path to the service entrance. A large current will surely blow the fuse or trip the circuit breaker (see Overload Protection, p. 37) and open the circuit. Even a small current not sufficient to blow the fuse will return via the ground wire and prevent a shock.

Note also that when we speak of the ground connection to the system neutral wire, we speak of "ground" or zero potential and we speak of "grounding" all boxes with a wire back to the service entrance. This last wire is often called a *bonding* wire to reduce confusion and differentiate it from the ground connections at the service entrance. Wherever the word "bonding" is used, remember that it means the wire that grounds the outlet boxes and so on. Ideally, that wire should never carry current but, if there is a defect, it will provide protection. It is better that an accidental current flow through the system than through you!

Just one more thing before we leave this subject. There is one exception to all of the above. The cases of appliances such as those shown in Figure 2-9 should not be grounded. Any appliance with heating elements placed where they could be reached, such as toasters or open-coil heaters, would become more dangerous with grounded cases. Think of someone poking a knife into a plugged-in toaster,

Fig. 2-9. If it ever becomes necessary to pry a slice of bread from a toaster, unplug the appliance first. (Graf–Whalen photo.)

to loosen a piece of bread (a dangerous practice under any condition!) and holding the toaster case to keep it from moving. If the knife touches the coils *and* the case were grounded, there would be a complete circuit through the knife, one arm, your body, the other arm, the case, the ground and the coils. Electrocution could be the result. It is better to keep the toaster case ungrounded to prevent this dangerous situation. But, in most other metal-encased appliances, where it will be difficult for the user to reach the electrical connections inside, the standard practice is to ground the case, thus placing a protective barrier between you and the hot wire should it accidentally contact the case (through insulation failure or breakdown).

Summing Up

Now that we have seen the many purposes that ground serves, let's review them one more time.

1. Grounding at the service entrance provides a safe method of ensuring that a failure in the power company's transformer will not result in overly high voltage being delivered to your house wiring.

2. Grounding also provides a safe path for lightning-induced voltages to be safely drained from power lines without causing insulation breakdowns and burn-outs.

3. In your home wiring system, grounding surrounds any and all electrical connection points with a conductive surface that is at the same neutral potential as the earth you stand on. Extending this to the case of a metal appliance means that there is no potential difference between it and other grounds you touch.

4. Grounding provides an enclosing, protective sheath so that a short circuit will merely result in the inconvenience of a tripped circuit breaker or blown fuse rather than exposing full line voltage on surfaces (such as switch plates, outlet plates, or other parts of your home wiring system). Not only does this promote user safety, but it also contains the showers of hot sparks that could cause a fire if a short circuit develops between a connection point and grounded box.

SHORT CIRCUITS

We discussed *short circuits* earlier, but the definition bears repeating. A short circuit means that an undesired circuit path develops, usually because of insulation breakdown, that bypasses an electrical component (such as a lamp or motor). The short circuit is a path of little or almost zero resistance, such as when the bared ends of hot and neutral wires touch together directly. This allows a large, uncontrolled surge of current that produces sparks, which can in turn start a fire. It is a situation that should be avoided. Fortunately, circuit breakers or fuses are part of each branch circuit serving your home and the main line feeding into your home. A short on a branch should quickly blow a fuse or trip the branch circuit breaker, thus limiting the possibility of fire.

OVERLOAD PROTECTION—FUSES AND CIRCUIT BREAKERS

Besides grounding, another safety factor that must be added to your system is overload protection. As you will learn to calculate later, each branch circuit of a house electrical system is designed to carry a certain number of lights and appliances that together require a certain maximum current. If the current rises above the designed limit, whether caused by a short circuit or by connecting too many lamps to the branch circuit, the conductors may overheat (remember, they have resistance, which means they can heat up if too much current flows through them) and create a fire hazard. For this reason, each circuit must be equipped with an overcurrent protector that will automatically interrupt (open or break) the circuit if this dangerous situation occurs. Most protectors are designed so that they will not cause "nuisance breaks" due to short-term current surges (such as when an appliance motor first turns on).

Overcurrent protectors can be either fuses or circuit breakers and must be the initial element in the hot line of every branch circuit of your system. For convenience, the fuses or circuit breakers of all circuits are normally housed in the service entrance panel at the point where electric power enters the home (see Fig. 2-10). Note that overcurrent protectors must always be placed in the hot line rather than the neutral. If the protector was in the neutral wire, even a "dead-short circuit" from the hot wire to a grounded box would not affect it. Large currents could, instead, flow through the grounding conductor (or an external ground) and a serious fire hazard could result.

The first fuse, made by Thomas A. Edison before 1900, consisted of a thin copper wire enclosed in the base of an electric lamp. He reasoned that this fuse

Fig. 2-10. At the service entrance panel each branch circuit feeds through an overload protector which limits the amount of current that can flow into the circuit. (Graf-Whalen photo.)

would be an intentionally weak link in the circuit and that it would melt and thus cut the circuit to protect the lamp in case of excessive voltage. Unfortunately, these early fuses were often as dangerous as the overload they were meant to protect against. Sometimes they would heat to incandescence and cause fires before they melted to break the circuit. Later fuse designs cured this problem by replacing the thin copper wire with a low melting point metal alloy.

In a standard plug-type fuse, current passes through the metal-alloy strip along the face of the plug. When the fuse blows due to overloading (too many lamps, heaters, or motor-driven devices, etc. are plugged-in), the metal strip overheats, melts, and breaks the circuit. When it blows because of a short circuit, the metal suddenly heats to such a high temperature that it vaporizes and discolors the fuse window.

Circuit breakers (see Fig. 2-11) are similar to fuses only in that they will break the circuit in case of overcurrent. However, they are not "kamikazes" that destroy themselves in the process. When they *trip*, they break the circuit due to a mechanical separation of contacts, much like those of a switch, rather than by the melting of a fusible link. After being tripped, a breaker can be reset by a switch lever and can be used over and over again. This same switch lever can also be used to de-energize a circuit at will, which is easier than removing a fuse and exactly the right action if you want to work on any part of the circuit. *Never work with an energized circuit.* Also, never try to cheat the system by pushing the lever on and holding it if it keeps tripping. When it trips repeatedly, it is telling you that there is a problem in the protected circuit and it must be fixed. (Fortunately, manufacturers realize that there are people who will try this anyway and all circuit breakers have been designed with this fact in mind. These breakers will trip

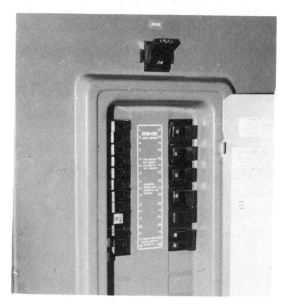

Fig. 2-11. A service entrance panel equipped with circuit breakers. (Graf–Whalen photo.)

regardless of whether the lever is held to the ON position.) Okay, so fuses and circuit breakers protect your house from fire. But, what protects *you* from electrical shock? The answer to this question is discussed in the next section.

GROUND-FAULT PROTECTION

A ground fault occurs when a current-carrying part of a circuit accidentally contacts any grounded, conductive material. In this case, current flows back to the grounded neutral connection by a path *other* than the neutral wire. Resistance of that path may be high or low, depending on the nature of the contact. Low resistance faults, such as contact with a grounded outlet box, draw heavy current. This will normally break the circuit through the over-current wires within a short time, almost instantly for a dead short circuit and up to a few seconds with less current.

However, when the ground fault has relatively high resistance, such as a human body if it accidentally gets into the circuit, the fault current is high enough to kill but not high enough to blow a fuse or trip a circuit breaker. Until relatively recently, there was no positive protection from this dangerous situation. However, since about 1965, protection has become available in the form of a life-saving device called the *Ground-Fault Circuit Interrupter* (GFCI), when used to protect people, or *Ground-Fault Interrupter* (GFI), when used as a system protector. When made part of a circuit, this ingenious little device constantly monitors the amount of current flow in both the hot and neutral conductors. As long as the currents in the two wires are identical, everything is fine. But, if there is a ground fault, current flowing to ground through a person or object not intended to carry it, the current through the hot wire will be greater than the current through the neutral wire. Even with one five-thousandth of an ampere (5 milliamperes) additional current, the GFCI's electronic circuitry can sense the difference. In effect, it knows that part of the hot line current is flowing to ground by other than the neutral wire and switches the electricity off within a few thousandths of a second.

Because fuses and circuit breakers perform an entirely different function than the GFCI, the GFCI is used in addition to the overcurrent protection device. To make this easier, there are replacement GFCI circuit breakers available to replace older, ordinary breakers that don't have the protection of the GFCI. We might describe the difference by saying that an ordinary circuit breaker only watches the current flow in the hot line and doesn't care how the current flows back to ground. The GFCI, however, monitors both the hot and neutral wires. In a circuit that is functioning properly, the hot and neutral currents must be identical. When there is the least difference in current (as little as 5 mA) bypassing the neutral wire and leaking back to ground (through you, perhaps), the GFCI senses the difference and opens the circuit. Circuit breaker and dual receptacle GFCI's are shown in Figure 2-12. Note that the dual receptacle GFCI is covered in Chapter VII along with instructions on its installation and wiring.

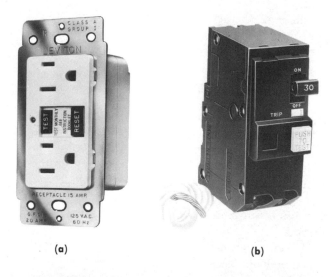

(a)   (b)

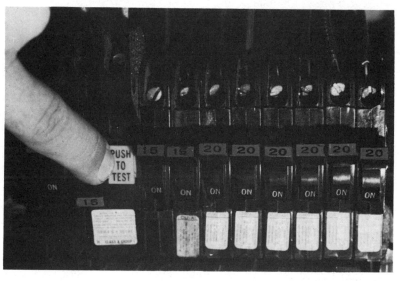

(c)

Fig. 2-12. (a) This dual receptacle GFCI can be installed in any outlet. It will afford protection for any cord plugged into its socket as well as for all other outlets that are beyond it on the branch circuit, i.e., are not between it and the service entrance. TEST and RESET buttons allow periodic confidence test that the GFCI is functioning properly. (Photo courtesy of Leviton Manufacturing Co., Inc.) (b) The circuit breaker shown combines the function of circuit breaker and GFCI in one easily installed assembly. This unit can be substituted for any circuit breaker of the same current rating. (Photo courtesy of Square D Co.) (c) When installed into the service entrance panel as shown, the circuit breaker provides double protection for the branch circuit in which it is installed. (Graf–Whalen photo.)

Remember that the phrase "current takes the path of least resistance" is not exactly true. Most of the current takes the easiest path, but some splits off and travels through any other parallel paths that are available. Although the bulk of the electricity may travel back to ground via the neutral wire, if a ground fault develops and your body becomes an accidental path to ground in addition to the neutral wire, some current, possibly lethal, will flow to ground through you. The extra current will hardly be noticed by the over-current device, but it will not fool the GFCI. The hot wire current will split into two paths and, because all of it will not return by the neutral, there will be a difference between the hot wire and neutral wire currents and the GFCI will spring into life-saving action!

Even with a GFCI, you may receive a very slight and brief shock in the event of a short circuit in the appliance you are using. But, because the GFCI very quickly senses the problem and shuts off the line current, you will feel only a momentary and very slight sting, which is good, because it warns you not to use the appliance again until it has been repaired!

The GFCI was originally developed for use in electrical circuits around swimming pools, where current leaks from the pump or water purification system could create an electrical hazard. It has special application in all wet areas, such as laundry rooms, but is equally useful with any appliance, such as a hair dryer, portable tools (drill, circular saw, etc.) or other devices you might use outdoors or where ground contact is likely.

However, even here the protection is not complete. If you accidentally connect yourself directly *across* the hot and neutral wires, even the GFCI will not help. There will be no difference in current in the hot and neutral wires and the GFCI cannot sense the danger.

So, even with all the built-in protective devices available today, you still need your common sense and strict vigilance. To finish this safety discussion, Tab. 2-1 contains some simple safety tips for working with and living with electricity.

Fig. 2-13. If a dark work area is not accessible to extension drop lights, use a good flashlight for your work. (Graf–Whalen photo.)

TABLE 2-1. The Cardinal Rules for Safety and Success in Doing Electrical Work

Never Work on an Electrical Circuit Without First Shutting Off Power at the Service Entrance Panel.

You need not shut off power to the entire house; just the circuit in which you are doing work. Also, be sure that everyone knows you are doing electrical work. The last thing you need is to have some annoyed family member reset a breaker on the circuit you are working. (If they do, it could very well be the *last* thing for you!) Remove and pocket the fuse or tape the circuit breaker in the OFF position. Close the fusebox or breaker box and tape a warning sign over it, so that the box cannot be opened without reading its message. Spell-out what you are doing and where you are. More than one weekend electrician has been jolted because this important step was overlooked. If possible, close and lock any door leading to the location of the box and be sure that everyone knows what you're doing.

Try not to work alone.

It is handy to have helpers within earshot if you need a tool or some other assistance. But, if you should accidentally contact a live circuit, that help could save your life. Think about this before you climb into a dark attic or concealed crawl-space to work on a wiring job. There is safety in numbers.

Only moles work in the dark.

What you cannot see clearly will be badly or dangerously done. Electrical work takes good craftsmanship and no craftsman works in the dark. If your job will take you into a darkened area, plan to have lots of light from battery-powered sources (see Fig. 2-13). Or, plug one or more extension drop lights into circuits far removed from the one on which you are working. Don't use dangerous sources such as candles or outdoor camping-type lanterns. Accidentally over-turning one of these in tight flammable surroundings could lead to trouble. Safe light sources are a must!

Plan to have the right tools before you start.

There's no easier way to make a wiring job a messy, wretched task than to start with the wrong tools! Every job involves some combination of tools, and you can size up your needs fairly quickly just by going over the area in which you'll be working, jotting down the steps, and then noting the kinds of tools you'll need. It's not difficult. Then, check your tool inventory and see if what you need is there. It may be that you'll have to purchase a tool or two. Or, perhaps visit a friend or neighbor with a "loan list." Either way, you win, because you'll have *what* you need, *when* you need it. Don't "make do"! You may wind up with a half-finished job at 11 o'clock on a Sunday night, with no possibility of turning the power back on, or worse, a make shift hack job that (1) isn't safe or (2) ruined the appearance of the area in which it was installed. The right tools make the difference!

Keep yourself out of the circuit.

Sure, you turned off the branch breaker or pulled the fuse before you got started, but has your work taken you back to some other junction box or connection point that may have a "live" circuit passing through it? If so, you will need to kill that one, too. Also, when you are working in an area through which several circuits pass, keep your body out of the circuits. How? Don't lean or kneel on metal conduit or armored cable (remember: these metal surfaces are grounded and so is the neutral wire) because you will be in contact with one wire of the ac line if you touch a grounded object. Likewise, don't drape yourself over pipes or other metal objects. Treat everything metal nearby as if it were a grounded object. Contact (even through clothes that are moist with perspiration) will ground your body, which means if you touch a "hot" wire, current will flow through you! So avoid grounds if possible. But certainly, shut off any circuit you even slightly suspect of sharing junction boxes with the one you are working on. *Keep yourself out of the circuit!*

3

SYSTEM DESIGN

Service Entrance – Branch Circuits – Calculations

As it should be, most of a household's wiring is completely hidden from view. Though surface wiring may sometimes be used and extension cords to lamps and appliances are common, the real wiring, whether armored or plastic-covered cable or conduit contained wires, is discreetly hidden behind plaster or wallboard walls, under floors, inside attics, and in basement ceilings. Because so much of it is out of sight, the home electrical system is somewhat of a mystery to most homeowners.

HOW YOUR HOUSE IS WIRED

In a typical three-bedroom, two-story house with attached garage, electricity usually comes into the house via overhead wires from the nearest pole of the electric utility that services the home. In most newer communities, the entrance wires are buried underground. Overhead or underground, the entrance wires lead to the main distribution panel for your system. This panel is generally called the service entrance and we will use that name throughout this book.

In most areas, power from the utility company pole is carried to the house by three heavy (thick) wires, which provide two different voltage levels, 115 and 230 V. Figure 3-1 shows a typical system service entrance for a three-wire power supply. The three-wire service cable from the utility company pole is twisted together with a steel support cable and clamped to the outside surface of the house at a point lower than the service entrance head. The three wires are then separated from the steel cable at the house. Each wire is formed into a drip loop as shown and enters the service entrance head, which is mounted higher on the

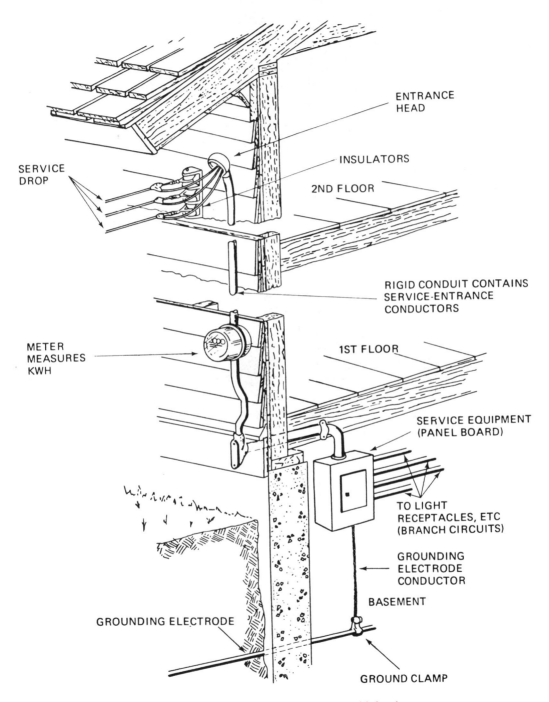

ENTRANCE
HEAD

INSULATORS

SERVICE
DROP

2ND FLOOR

RIGID CONDUIT CONTAINS
SERVICE-ENTRANCE
CONDUCTORS

METER
MEASURES
KWH

1ST FLOOR

SERVICE EQUIPMENT
(PANEL BOARD)

TO LIGHT
RECEPTACLES, ETC
(BRANCH CIRCUITS)

GROUNDING
ELECTRODE
CONDUCTOR

BASEMENT

GROUNDING ELECTRODE

GROUND CLAMP

Fig. 3-1(a). Typical three-wire system entrance with fuse box.

Fig. 3-1(b). The service head on the home, to which the power company's service connects.

Fig. 3-1(c). The electric meter is serviced and read by the utility company. (Graf–Whalen photos.)

wall than the cable support so that rain water will drip off rather than enter the service entrance head. The three individual wires from the utility pole, two hot and one neutral, are spliced to three wires of the entrance cable, which runs down the wall to the meter box (this cable contains a red, a black, and a white wire). [Forget the meter for the moment; your utility company will install and maintain it.] At the other side of the meter box, a second length of cable, identical to the cable down the wall, carries power into the house through the wall (unless, as is often the case in older installations, your meter is inside the house).

This second cable leads to the service entrance, a main distribution panel containing system overcurrent protection and shutoff devices (fuses or circuit breakers sized to handle the entire system input current) and the terminations of all the system branch circuits. The branch circuits split up the incoming power into small, carefully-designed fractions of the total input power. Each of the branch circuits uses lighter (thinner) wire than the main input cable (see Fig. 3-2a) because they carry only a fraction of the input current. At the panel branch circuit terminations, therefore, overcurrent protectors of lower current rating than the main device are installed. These overcurrent protectors (see Fig. 3-2b) are sized to protect the circuit in case current exceeds the limit that the branch circuit wire can safely carry. You could use very heavy wiring for a branch circuit but that would be wasteful and impractical. Branches are designed to carry lower current and wired with the right size wire to safely carry that current. The service entrance (including ground connection) and branch circuit connections are described in detail in Chapter 6, but here we need just see that the power is divided among the branch circuits.

Fig. 3-2(a). The main input cable is sized to carry current for all the branch circuits. (Graf–Whalen photo.)

Fig. 3-2(b). Circuit breakers sized by amperage rating protect each branch circuit from excessive current. In the panel shown, one circuit breaker is also a GFCI. (Graf–Whalen photo.)

Figure 3-3 is a schematic representation of how the power divides. Note that from the three-wire input, circuits having two different voltages are derived. Some circuits connect from a hot wire (black or red) to neutral wire (white), and some connect across the black to red wires. From each hot wire (black or red) to neutral, the utility company supplies 115 volts, providing two different groups of 115-volt circuits. Furthermore, these voltages are additive, so that when you connect from one hot wire (black, for example) to the other (red), you get a total of 230 volts. The 230-volt circuits provide heavy-duty, maximum-wattage power but allow wiring to be of reasonable size, because doubling the voltage allows the current to stay at the same level it would be in a 115-volt circuit while delivering twice the power.

The 115-volt branch circuits are used for general purpose, normally light-duty service, for lamps, TV, vacuum cleaners and the like. Overcurrent protectors rated at 15 to 20 amperes are installed in the hot line only, as indicated in Figure 3-3. With 15- or 20-ampere capacity and 115 volts, the most power available is 1725 to 2300 watts (15 A x 115 V or 20 A x 115 V), but this can operate quite a few lights and other 115-volt devices. The 230-volt circuits have overcurrent protectors in each hot line, and a 20-ampere capacity circuit can deliver 4600 watts (20 A x 230 V). These figures help to explain why large loads, such as air conditioners, electric stoves, and electric water or space heating equipment, are usually operated from the 230-volt circuit.

One other circuit component, shown in both Figures 3-1 and 3-3, is the ground connection. The neutral wire of the incoming three-wire cable is con-

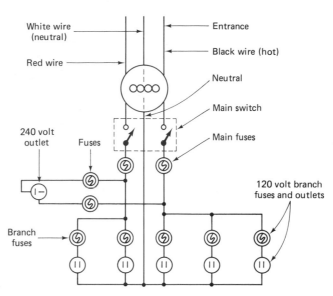

Fig. 3-3. Schematic representation of how the power divides.

nected to earth ground outside the house (see Chapter 6 for details). Within the service entrance panel, this ground is connected to the common bus, to which the neutral (white) wire of every branch circuit is also connected. This bus provides the ground connection for each circuit through a separate green or bare wire, so that all outlets, receptacles, fixtures, and so on can be bonded to ground for safety, as described in Chapter 2. The equipment ground (green or bare wire) must be connected to the ground bus. Where branch circuits originate at the service entrance panel, the neutral bus (which is grounded) can be used for terminating green or bare connectors.

In most areas of the country, electrical power is most commonly supplied by the three-wire system we have been discussing. However, two other types are also in use, namely two-wire and four-wire systems.

The Two-Wire System

Electric supplies for most houses built prior to World War II were of the two-wire variety and limited to 115 volts, generally with a maximum service of 60 amperes. These systems can only supply a total of 6900 watts and are considered inadequate by present standards, so inadequate, in fact, that we hardly discuss them. You could replace an outlet or switch, but that is probably the limit of what can be done to improve with this kind of wiring system. Also, if you install a new service entrance to expand your household power capacity, you certainly would not want another two-wire system unless upgrading was absolutely impractical.

The Four-Wire System

This system is typically used for power service to factories, apartment complexes, large stores, office buildings, and other high power consumption industrial/commercial applications. It is only occasionally used for household service but it is described here for the sake of completeness. The a-c cyclic waveform we discussed in Chapter 1 is the waveform of ac from a simple, basic generator. This is called single-phase power because only one power peak occurs in each polarity during the 1/60 second of a complete cycle. In actual practice, power stations do not generate single-phase power. Instead, their alternators develop *polyphase,* or *three-phase power,* and single-phase circuits are derived from it as needed.

Figure 3-4 shows a three-phase waveform as it would be generated from a modern alternator. Curves A, B, and C are separate voltage waves that flow in the same circuit 1/180 second apart. The three windings on the alternator are, in effect, like the outputs of three separate generators. As the three-phase power reaches the distribution point, three-phase power is delivered in a four-wire cable from the utility company transformer.

The arrangement is shown on Fig. 3-5 and is actually simple enough. One of the four wires is white and is the system neutral wire. It is grounded at the service entrance. The other three wires are red-covered and are each hot wires. Between the ground wire and each of the red wires, there is 115-volt single-phase power for the usual 115-volt household machines and appliances. Each branch circuit has its own fuse and these branch circuits can be treated just like any other 115-volt circuit from a two-wire or three-wire system. However, the voltage between any two red wires of a four-wire circuit is 208 volts (not 230 as you might expect from our discussion of three-wire circuits). This difference has to do with the geometry of the alternator and is quite involved.

Note: *If the voltage between phases is 208 volt the voltage between any phase and neutral must be 120 volt.*

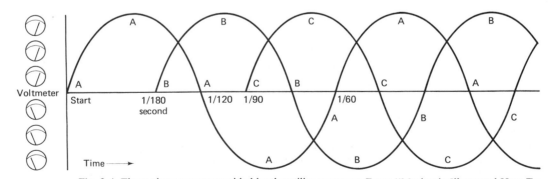

Fig. 3-4. Three-phase power provided by the utility company. From "Mechanix Illustrated How-To-Do-It Encyclopedia," copyright 1961, Fawcett Publications, Inc.

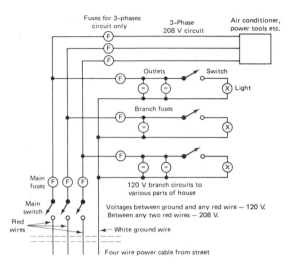

Fig. 3-5. Four-wire, three-phase power system. From "Mechanix Illustrated How-To-Do-It Encyclopedia," copyright 1961, Fawcett Publications, Inc.

Also, connecting heavy three-phase loads, such as a large air-conditioner motor, requires an intricate interweaving of the three phases and is beyond the scope of this book. We must therefore drop this discussion of four-wire circuit and assume that you will only be using this book for three-wire, 115/230-volt electrical installations.

PLANNING YOUR SYSTEM

Service entrance equipment is rated in current-carrying capacity (ampacity, measured in amperes) and must have sufficient capacity to accommodate the maximum amount of current that might logically be carried at any one time. Generally, 230-volt, 100-amp service, which may be provided by No. 3 AWG conductors with RHW insulation and a 100 A/230 V capacity service entrance panel is the minimum recommended for new homes. 150- or 200-ampere service, using No. 2/0 or 1/0 wires, would be even better, and you should consider carefully what you might want in the future before you lock yourself into a system with inadequate capacity. (Wire sizing and insulation are discussed in detail in Chapter 4.)

In planning a complete house system (e.g., for a new house), your first step should be to consult your local power company. Do they supply three-wire 115/230-volt service? Can you get 100-amp or 200-amp service if you want it? Of course, you might just get by with 60-ampere service, but nowadays 100-ampere, three-wire service is the minimum you should install in any house of up to 3000 square feet in floor area. Better still, select the largest capacity that will fit into your budget. With a 200-ampere service you will not likely be limited later in your choice of appliances, so if at all possible, try to go for it.

Your power company can also provide other important details to aid your planning. For example, where will the service entrance be located? A particular entrance route may be necessary to avoid having wires pass over other people's property or to avoid low-hanging wires over a highway. If your house is a one-story structure, it may be necessary to install an entrance mast to hold the wires high above the ground. Because the location of the service entrance will determine the general layout of your inside wiring, this is an important detail to nail down and your utility company representative is the person to ask.

The power company can also give you information on the amount of work the company's men will do in connection with the hook-up. This may vary over a wide range. In some areas, the homeowner is responsible for all the wiring from the service entrance panel to the meter box socket and from the meter socket up to the entrance head. Then, the power company will install and connect its wires to the service head, install the meter, and the system is ready for inspection and use. In other areas, you may be responsible for wiring beyond the service entrance head to the pole and from pole-to-pole if several are needed to reach the distribution point. The power company may simply install the meter and make the final connection to its transformer. In any case, find out first.

Let's assume that you are satisfied with the location of the entrance, what the power company will do, and what service will be available. Now comes the real planning. First, what amperage service will you choose? Then, how many circuits will you divide it among; how many will be 115-volt and how many 230-volt circuits? On each circuit, how many outlets will you need; how many outlets will you want in each room?

Your copy of the National Electrical Code can be your most helpful guide in answering these questions. To comply with the code, there must be enough lighting circuits to supply 3 watts for every square foot of floor space in your house. This figure is easy to compute. Spacing the outlets is another problem and again the code has a guide. Install outlets "so that no point along the floor line in any useable wall space is more than 6 feet, measured horizontally, from an outlet." In effect, this says your outlets should not be more than 12 feet apart in any room. The distance from a door opening to an outlet cannot be more than 6 feet measured along the wall.

Lighting circuits are general-purpose (15-ampere) circuits for operating lamps and small appliances such as fans, vacuum cleaners, and TV sets. Special appliance circuits are needed for heavier loads such as toasters and electric irons; these, generally, have 20-ampere capacity. For major appliances, electric kitchen ranges, water heaters, space heaters and clothes dryers, you need individual, dedicated circuits, usually 230-volts for each. The greater the power required by an appliance, the greater the current capacity the circuit will need. This, in turn, determines the size of the wiring to be used, which is discussed in more detail in Chapter 4.

An important consideration in your planning is to decide what tools and appliances you will need in your home now and in the future. It is much easier to install the wiring and circuits for them while the overall system is being installed rather than waiting until later. Also, this decision helps determine the service capacity and, therefore, the capacity of the service entrance panel that you will need. Table 3-1 contains a listing of the power required for typical electrical devices and can be used to add up your estimate of the power you need.

Calculations

Given below is a typical calculation for a single-family dwelling electric system design. This example is derived from similar examples given in the National Electrical Code which you should have handy for any planning you do (see Fig. 3-6). We have taken the liberty of simplifying the calculations but the example should be sufficient to show the principles involved. (Note that the examples in the NEC use 115/230 volts for calculations, because 115/230 volts is the average voltage level throughout the U. S. power system. In practice, 120/240 volts is the level your power company attempts to maintain.)

EXAMPLE: A dwelling has a floor area of 1500 square feet exclusive of un-occupied cellar, unfinished attic, and open porches. In addition, it will have a 12kw (12,000 watt) electric range. Compute the ampere capacity (ampacity) of the conductors that will supply the service entrance. Circled numbers refer to notes at the end of this example that explain the NEC requirement.

Fig. 3-6. At the planning stage, the NEC should be consulted. (Graf–Whalen photo.)

TABLE 3-1. Power Requirements of Typical* Electrical Devices

Appliance	Average Wattage Rating
Blender	390
Broiler	1,450
Carving Knife	100
Coffee maker	900
Deep fryer	1,450
Dishwasher	1,200
Egg cooker	500
Frying pan	1,200
Hot plate	1,250
Mixer	125
Oven, microwave	1,450
Range with oven	12,200
Roaster	1,300
Sandwich grill	1,150
Toaster	1,150
Trash compactor	400
Waffle iron	1,100
Waste disposer	440
Freezer	
(14 cu. ft.)	340
(frostless - 15 cu. ft.)	440
Refrigerator	
(12 cu. ft.)	300
(frostless - 12 cu. ft.)	390
Refrigerator/Freezer	
(14 cu. ft.)	325
(frostless - 14 cu. ft.)	600
Clothes dryer	4,850
Iron (hand)	1,000
Washing machine	
(automatic)	500
(nonautomatic)	280
Water Heater	
standard	2,475
quick recovery	4,475
Water pump	460
Air cleaner	50
Air conditioners—room	1,500
Bed covering	175
Dehumidifier	250
Fans	
attic	370
furnace	290
window	200

TABLE 3-1. (Continued.)

Appliance	Average Wattage Rating
Heater (portable).............................	1,320
Heating pad	65
Humidifier	175
Oil burner or stoker............................	265
Radio	70
Radio/record player	100
Television:	
black and white	
tube type	160
solid state..............................	55
color TV:	
tube type	300
solid state	200
Clock	2
Floor polisher	300
Sewing machine................................	75
Vacuum cleaner	630
Electric brooder	100 (and higher)
Milking machine...............................	250 (and higher)
Milk cooler	500 (and higher)
Milk pump	200 (and higher)
Barn fan	125 (and higher)
Barn cleaner	1,500 (and higher)
Feed conveyor.................................	375 (and higher)
Elevator	375 (and higher)

*Remember that these are averages for determining rough estimates. Check the actual ratings of your own home inventory of products if you are at all in doubt about a circuit's capacity.

Computed Load
1. General Lighting Load:

   1500 square feet at 3 watts per square foot = 4,500 watts.[1]

Minimum Number of Branch Circuits Required
1. To Handle General Lighting Load:

$$\frac{4,500\text{W}}{115\text{V}} = 39.1\,\text{A} \simeq 39\,\text{A (rounded off)}$$

   Either three 15-ampere two-wire circuits or two 20-ampere two-wire circuits can cover the above requirement.
2. Small Appliance Load:

   Two 20-ampere, two-wire circuits[2]

55

3. Laundry Load:

One 20-ampere, two-wire circuit[3]

Minimum Size Feeders Required
Computed Load:

General Lighting	4,500 watts (above)
Small Appliance Load	3,000 watts[4]
Laundry	1,500 watts[4]
Total (without range)	9,000 watts
Computed load is therefore:	
3000 watts at 100%	3,000 watts[5]
9000-3000 = 6000 watts at 35% =	2,100 watts[5]
Net Computed Load (without range)	5,100 watts
Range Load (12,000 watts at 67%)	8,000 watts[6]
Net Computed Load (with range)	13,000 watts

For 115/230-volt, 3-wire system feeders

$$\frac{13,000W}{230 \ V} = 57 \ A$$

The Net Computed Load exceeds 10,000 watts, so service entrance conductors must have 100 amperes ampacity.[7]

Notes on NEC Requirements

1. General lighting circuits for dwelling units shall supply a minimum of three watts per square foot of dwelling space.

2. In addition to general lighting circuits, no less than two 20-ampere small appliance circuits shall be provided for small appliance loads, including refrigeration in kitchen, breakfast room, and so on.

3. At least one additional 20-ampere branch circuit shall be provided to supply the laundry receptacle outlets. This circuit shall have no other outlets.

4. The feeder load shall be computed at 1500 watts for each two-wire 20-ampere small appliance circuit and each two-wire 20-ampere laundry circuit.

5. The first 3,000 watts of lighting loads shall be computed at 100% demand factor, from 3,000 to 120,000 watts at 35% demand factor.

6. Demand factors for household electric ranges and similar equipment are computed in accordance with a table in the NEC. For 12,000 watts, the demand factor is 67%.

> **Note:** *Demand factor is a measure of the probable use of a circuit; it is the ratio of the maximum demand to the total load on the system and is measured in percent.*

7. Ungrounded, service entrance conductors shall be rated at 100 amperes, three-wire for a dwelling unit with six or more two-wire branch circuits or a net computed load of 10 kilowatts or more.

CIRCUIT LAYOUT

Now you know where the service entrance will be, what the utility company will do for you, what ampere service is required, and how to calculate the minimum number of branch circuits you will need. This must all be pulled together in your final layout. You must plan your wiring routes, locate all outlets, ceiling fixtures, and switches, decide how many branch circuits you would like (preferably much more than the minimum), and how many outlets you want on each branch.

A typical modern system will probably have five general purpose circuits, three kitchen-dining and laundry room circuits, and ten or more special purpose circuits for individual pieces of equipment. The NEC does not specify the maximum number of outlets you may include in a circuit; this is simply a matter of convenience for furniture layout and so on. To avoid trouble, you should determine how many outlets are likely to be in use at one time (you could call this the demand factor) and whether so many will be used that an overcurrent protector will react.

Also keep in mind that every circuit has a maximum wattage capacity. A 15-ampere circuit has a capacity of 1,725 watts (15 A x 115 V) and a 20-ampere circuit, 2,300 watts. It makes good sense to limit these maximum loads, for example, to 1,440 watts for 15-ampere and 1920 watts for 20-ampere ciruits. These limitations point out the value of adding more circuits than the minimum number you might arrive at by the NEC calculation. This extra margin is particularly helpful whenever a motor may be plugged into the outlet, because motors draw particularly heavy currents during the first seconds of startup. It is better to derate the circuit (reduce your capacity expectation) to less than 80% when refrigerators or other motor loads of this sort will be on the line. In accordance with the NEC, appliance circuits must be separate from general purpose lighting circuits, but they are limited in capacity by the same formula. Kitchen gadgets come in a great variety of types and use more watts than you might imagine, so you must be particularly careful to provide enough capacity in that area.

Figure 3-7 shows a typical house floor plan, marked with circuit routes and devices. A good way to key your plan is to number the circuits and put a small corresponding number alongside each outlet, switch, or fixture on the circuit. For clarity, you might also draw separate plans with all general-purpose circuits on one page, kitchen and laundry on another, and 115-volt or 230-volt circuits to individual large-current loads on a third. The circuit wiring routes could be differentiated by different colors or numbers along each line. Three dimensional skeleton views of the house from basement to attic, with the locations of devices and wiring routes related to floor levels as well as wall perimeters, would be an even better model with which to work.

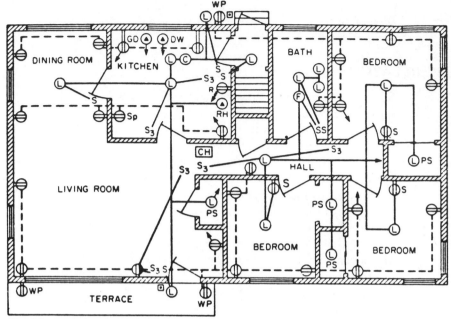

Fig. 3-7. Typical residential floor plan shows outlets spaced along all interior walls (12-foot maximum spacing between them). Switches (S) controlling ceiling fixtures in dining room, bedrooms, and bath; a set of three-way switches (S-3) controlling the ceiling fixtures in the kitchen; pull chain lights (PS) in all the closets, and weatherproof (WP) outlets outside the front and back doors. Lines between outlets and switches indicate the branch circuits, and arrowheads show the cable end leading to the service entrance box. You can check the number of circuits in this plan by counting the arrowheads. From Doyle, John M., *An Introduction to Electrical Wiring,* © 1975. Reprinted with permission of Reston Publishing Company, Inc., a Prentice-Hall Company, 11480 Sunset Hills Road, Reston, VA.

In addition to the NEC requirements for receptacle outlets per wall perimeter, which we mentioned previously, there are many other considerations to keep in mind when planning your wiring layout. The following list covers most of the important points.

1. General purpose circuits must have an outlet receptacle for every 12 feet of wall perimeter throughout the house.

2. Outlets are placed only on those walls two feet wide, or more.

3. Install receptacle outlets in the wall 18 inches above the floor.

4. Outlets in kitchen and dining areas should be equally divided between the two required 20-ampere circuits.

5. Install kitchen and work area receptacle outlets 8 inches above the counter or work table (see Fig. 3-8b). Check this requirement against your local code before installation.

6. Install one dual receptacle for each 4-feet of kitchen counter length.

7. All outlet boxes are to be connected to the entrance ground. In electrical parlance they must be "grounded".

8. Weatherproof outlets are desirable outdoors on the driveway side of the house and the side opposite the driveway. Place these outlets about 4 feet above ground. Weatherproof light fixtures should be mounted above or at the sides of all exterior doors. At least one outlet must be installed outdoors for all single-family residences.

9. All circuits serving outdoor outlets or fixtures must be protected by ground-fault interrupters.

10. Have at least one wall outlet in the dining room above table-height.

11. Place outlets for most ceiling lights in the center of the room. Place bathroom lights above mirrored cabinets. Receptacle outlets for electric shavers and the like can be built into this same fixture. All outlets in the bathroom must be protected by a GFCI.

12. Position boxes for switch controls about 48 to 54 inches above floor level at the latch side of the doorway (see Fig. 3-7).

13. Consider three-way and four-way switch controls to operate lights from more than one location, for example, at top and bottom of stairs or at both ends of a long corridor (see Chapter 8).

14. Arrange for at least one receptacle outlet for living room lamps to be switch-controlled from the doorway.

15. Include lights in closets wherever possible. These may be controlled most easily by a pull-chain in walk-in closets. Recessed fixtures with solid lenses have to be used in closets that do not provide 18 inches of clear space to combustible materials.

16. An Alabax porcelain vapor-proof shower light, such as that made by Pass & Seymour, is a decided convenience. Control it by a wall switch not accessible from the bath.

17. Install light switches with illuminated switch handles, if possible.

18. Appliances that are heavy users of electricity are given their own fused circuits. Extra heavy users need a 230-volt supply with an appropriate overcurrent protector in each hot leg.

19. Some appliances may be designed to use both 115-volt and 230-volt current. Many electric ranges, for example, operate on 230-volt for high heat and 115-volt, for low heat.

20. Select copper conductors for your circuit wiring, with ampere capacity ratings to match the overcurrent device for each circuit. See Chapter 4 for selection criteria.

21. Voltage drop can be a problem in long wiring runs. For example, a 70-foot-long house needs a larger-than-normal conductor for a run from the service entrance at one corner, to the attic at the other. See Chapter 4 for design criteria.

OLD WORK

We shouldn't leave this subject without some discussion of *old work*. In general, the problem with existing systems will be how to add capacity. If fuses blow too frequently or you want to add a new large piece of equipment such as an electric range or central air conditioner, new circuits are the remedy. You can add new outlets on old circuits, but because they only add convenience, not capacity, you probably won't have need for them often. Also, this may only be done if the existing outlets are on a grounded branch circuit (see Fig. 3-8). The Code forbids installation of any ungrounded outlets, as well as installation of grounded outlets on old, ungrounded circuits (see Chapter 7 for outlet installation). Check your service entrance ground connection (see Chapter 5) to determine proper installation. Also, if you add to an existing circuit, you must use wire at least as large as that on the existing circuit (see Fig. 3-9).

Fig. 3.8. There is no more room for addition of circuits in this panel. Replacement with a new panel is recommended. (Graf–Whalen photo.)

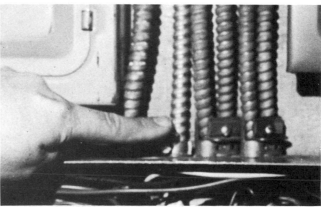

Fig. 3-9. An addition to an existing circuit requires wire as large as that already on the circuit. (Graf–Whalen photo.)

As far as increasing your system capacity is concerned, your first task will be to check carefully what you have. Inspect all visible wiring in outlet boxes for evidence of insulation deterioration or for errors the original installer may have made. Check Chapter 7 for proper color coding of wiring and see that your system is properly wired before proceeding. If anything is wrong, you might first have to replace old wiring before expanding on the system. Remember, the old system is your "foundation". If it's in bad shape, it won't support your new work, no matter how good a job you do.

Next, draw the layout of your existing system. You can usually do this by plugging a test lamp into outlets and removing fuses until you see which one turns out the lamp. Wall and ceiling fixtures can be checked the same way, and switches can be checked simply by operating them. You should tape a copy of this information to your fuse box for quick reference.

When you have finished this survey, work backwards using the information in Chapter 5 and determine whether your service entrance equipment can handle added circuits. This is not always possible. You may find that in order to add capacity you will have to add an entirely new service entrance. In that case, you can go back and use the information given in this chapter for design (if you do add capacity this way, we urge you to include room for a large expansion) and use the information in Chapter 5 for the actual work.

4

HARDWARE

Wire, Cable, Conduit – Boxes – Tools

A successful electrical wiring installation requires good quality hardware. You should become familiar with all types of hardware available on the market today, such as the service entrance distribution box, fuses and circuit breakers, electrical cables and conduit to outlet boxes, switches, lighting fixtures, and so forth. You will also need a complement of basic tools for bending, shaping, and connecting wires and for installing boxes and fixtures as well as for carpentry, wall repair, and other tasks needed to do a good job.

This chapter gives an overview, describing major types of each item. The available variety within each category is very great. However, you won't need everything, and just knowing what each item looks like and what it does will enable you to quickly focus on the items you need. Just remember, though, that you must have a copy of the local electrical code to see what type of hardware is allowed or required for specific jobs. Then (if it applies to the item you need) insist on UL-listed products and you will be off and running. The basic hardware parts of your system, which we will discuss in order, are: wires; cables; conduit and wire connectors (solder and solderless); surface wiring; and extension cords to lamps and appliances. Next come the entrance panel (fuse or circuit breaker type); fuses and circuit breakers; switches and fixtures; boxes and hangers for boxes. The tools, both a basic complement that you must have, and a suggestion for short-term rental as needed, are the final points covered in this chapter.

CONDUCTORS AND CABLE

Your home electrical system comprises many separate elements distributed throughout the house. These are connected together and integrated into one system by means of electrical conductors and you should be familiar with the types that are available.

As we mentioned in Chapter 1, an electrical conductor is a substance that conducts or allows transmission of electricity as opposed to an insulator, which does not. Conductors for residential wiring are almost always made from copper. Aluminum and copper-clad conductors are offered in some localities, but these have special limitations and should be passed over, in favor of copper. Conductors are either made of single, solid wire or composed of many strands of thinner wires that are twisted together. Such wires are called *stranded* wire and they are generally covered by a close-fitting coating of rubber or plastic insulation. Uninsulated, bare, solid wire conductors are often provided for grounding purposes only. Also available are insulated conductors wrapped together in an overall insulating jacket to form two- or three-wire cables.

The current carrying capacity of the wires, also called *ampacity,* depends on the material and the size of the wire. Copper wires can carry much more current than aluminum wires of equal size. Therefore, copper conductors are generally preferred. Also, copper can be soldered easily, whereas aluminum requires special techniques. Although aluminum may be shaped more easily than copper, it has a tendency to "creep" and, with age, can work loose from a screw terminal. Copper, once formed and fastened, stays put much better. Also, copper will grip tightly in a push-type terminal, whereas aluminum must be fastened by a screw terminal. For a given ampacity the NEC requires that, if you use aluminum wire, it must be *two sizes larger* than a comparable copper wire. This can cause spacing problems in a tight junction box. All things considered, copper is the better choice, though it is generally more expensive.

To further support our suggestion that you use copper conductors, consider this. When aluminum wiring first came into use years ago, it was connected to terminals (or switches, outlets, and the like) that had been designed for copper wires. Because problems caused by aluminum's aging were unkown then, connections would loosen and heat badly. In many cases, fires and other safety hazards resulted. To eliminate these problems, terminals, switches, and similar devices were redesigned at UL's insistence. Subsequently, components considered suitable for use with copper or aluminum were marked CO/AL. This, too, had to change when it was later found that the 15- and 20-amp devices (the most frequently-used sizes) were still not suitable for use with aluminum wire and had to be further redesigned. Now, *only* devices marked CO/ALR are acceptable, for units rated below 20 amperes, while devices marked CO/AL are still good for units over 20 amperes.

UL also requires that the markings CO/AL and CO/ALR must be stamped into switch mounting yokes, receptacles, and other units so that they are visible

without removal from the boxes in which they are installed (see Fig. 4-1). If your present system was installed some years ago and employs aluminum conductors, survey it to be sure that CO/AL or CO/ALR receptacles and switches are used throughout. If not, replace all unacceptable units with properly marked devices for safety's sake.

A compromise conductor is also available, called copper-clad aluminum. This hybrid wire may be used with any terminal design, including push-in types. This does not solve the ampacity problem, so, for comparable current carrying capacity you must still use copper-clad aluminum that is two sizes larger than copper.

For all the reasons given above, copper is still the best choice for a conductor. When we speak of conductors, we will therefore always mean copper unless noted otherwise.

Wire Size Is Important

You can see that conductor *material* is important. If you decide to use only copper conductors, you need to know only a little more. You will often use different size conductors for different applications, so let's see why the size of the wire is so important.

0   1   2   4   6   8   10   12   14

Fig. 4-2. Actual cross-sectional areas of copper conductors in common use. (Courtesy of Sears, Roebuck and Co.)

TABLE 4-1. AMPACITY

ALLOWABLE CURRENT-CARRYING CAPACITY, IN AMPERES
OF COPPER CONDUCTORS*
Based on room temperature of 30°C (86°F)

Size of Conductor AWG	IN RACEWAY OR CABLE Not more than three conductors in raceway or cable; if the number of conductors is four, the allowable carrying capacity is 80 percent of the values given.		IN FREE AIR	
	Rubber Types R, RW, RU, RUW ——— Thermoplastic Types T, TW	Rubber Type RH	Rubber Types R, RW RU, RUW ——— Thermoplastic Types T, TW	Rubber Type RH
14	15	15	20	20
12	20	20	25	25
10	30	30	40	40
8	40	45	55	65
6	55	65	80	95
4	70	85	105	125
2	95	115	140	170
1	110	130	165	195

Conductor sizes are designated by a number. In the U.S., the American Wire Gauge (AWG) system is used and wire you purchase will be identified as No. 14 AWG, No. 12 AWG, and so on. Figure 4-2 shows actual cross-sectional areas of the most common range of conductor sizes in current use. These cross-sections show the metal conductor material itself without any insulation covering.

The larger the conductor size, the smaller the AWG number. Sizes 1/0 and 2/0 really mean size 0 and 00; even heavier sizes 3/0 and 4/0 are available, 4/0 being nearly a half-inch in diameter. Wires beyond this size and up to several inches in diameter use a different numbering system. Note that the heavier wires shown are made of stranded conductors—many smaller conductors twisted together. Stranding is used to provide better flexibility. This cuts down somewhat on the active cross-sectional area of the wire, so these wires are thicker than a single solid conductor of equal ampacity would be.

No. 14 wire, fairly thin and easily formed, is the size most often used for branch circuit residential wiring, though wire up to No. 1 may be used in the main cable going into the service entrance. Table 4-1 shows the ampacities of various sizes of copper and aluminum wires. No. 16 and No. 18 wires are used mostly in flexible cords, while still smaller sizes are used in the construction of electrical motors and other electrical equipment. No. 18 may also be used in low voltage (under 30 V) wiring systems that operate doorbells, chimes, thermostats, and similar equipment, as described in Chapter 10. We are not through yet. If you decide to use No. 14 wire for a 15-ampere branch and No. 12 wire for a 20-ampere branch, you will normally be in good shape. However, suppose the load is located at a distance, for example, one hundred feet or more, from the service entrance. Now, voltage drop due to resistance comes into play.

Remember that in Chapter I, we said that current flowing through a conductor encounters some opposition and thus wastes energy in generating a certain amount of heat. The greater the current, the greater the loss through heating. And, the longer the distance, the greater is the resistance, so power loss in conductor heating becomes appreciable.

This loss would be nuisance enough if it only created a small loss of power. It does more, though, because it also causes the voltage to drop from what is available at the beginning to what is left at the end of the wires—and this can mean a serious efficiency problem. If No. 14 wire is used, 115-volt service will drop to only 114 volts 10 feet away and to about 108 volts 100 feet away, with a current of just 2 amperes flowing in the circuit. Electric appliances work inefficiently on voltages lower than their design rating. A motor, for example, will only produce 81% of its normal mechanical power when the voltage drops to 90% of its rated value.

As you can see, conductor resistance is a key factor in getting best performance from your appliances. A larger diameter wire has lower resistance. And so, if you spend a little extra money in your initial installation by choosing larger wire, you won't suffer year after year in lost efficiency of your appliances. Use No. 12 wire for 15-ampere branches, No. 10 for 20-ampere branches, and so on. These will service the furthest points of a large house or outbuildings and swimming pools at any reasonable distance from the house with much less voltage drop than if lighter wiring were used.

Wire Insulation and Cabling

Of the many types of wire manufactured today, only a few are commonly used in the home. All must be suitably insulated for up to 600-volt service, except for those wires intended for low voltage circuits, such as doorbell circuits (Chapter 10) and fixture wires (300 volts). Specific insulated wire types must also be used for specific applications; some are designed for use in damp locations, such as underground and outdoor wiring, while others can be used only in dry locations inside the house. Figure 4-3 shows common types of cable and their application limits.

Because we are only talking of copper conductors, the principal factor determining which wire to use for a given application is the method of insulating the wires and of jacketing groups of wires into cable.

Conductors for residential use are generally made today with a thermoplastic insulation covering. The thickness of the insulation depends on the size of the wire. The insulation is also manufactured in different colors, such as black, red and white, or grey, to allow easy identification of the same wires at both ends of a cable or conduit run. This coloring is called *color-coding*.

Service Entrance Cable for supply to the main distribution panel from the meter. It is usually a three-wire cable, the third wire being a bare grounding conductor.

Romex or NM Cable for indoor use. Available in three-wire and four-wire types. The three-wire cable is called two-wire with ground; and the four-wire cable is called three-wire with grounds in this book.

Underground or Type UF Cable for outdoor application. The jacket is generally a molded plastic for moisture proofing. It is usually available in two-wire with ground configuration.

Armored or BX Cable is used for the same application as Romex. It is extremely durable but slightly harder to work with than Romex. It can be used only in dry, indoor circuits.

Fig. 4-3. The most popular types of available cable configurations. (Courtesy of General Switch Company, Middletown, N.Y.)

The most common modern wire that may be used in either wet or dry locations is called type TW wire. Type THW wire is similar, but it is designed to withstand a higher degree of heat. In older homes, you will probably find rubber-insulated wires, but these are no longer considered satisfactory because the rubber deteriorates, drys-out, and breaks over years of exposure to oxygen and ozone. Other modern wire types are designated THHN and THWN. Their insulation is a combination of thermoplastic covered with an extruded jacket of nylon. For a given conductor size, these wires are a bit thinner than others and are better for use in conduit runs. A still more recent type of insulated wire is called XHHW and it uses an even thinner insulator, made of cross-linked synthetic polymer. This material has superior insulation value and heat resistance.

Two or more wires grouped together and covered by an additional outer jacket are called *cables*. Where permitted by the local code, cables present the most convenient method of house wiring. Metallic sheathed cable, called BX*or armored cable, was the industry mainstay for a long time, but lately, cables with nonmetallic covering have been gaining widespread popularity. A cable with two No. 14 wires jacketed together is called 14-2. Two No. 14 wires plus a ground wire is 14-2 with ground (*not* 14-3). Three No. 14 wires in a cable would be 14-3. This same numbering arrangement naturally applies for other wire sizes.

In a two-wire cable, one wire will be white (neutral conductor) and the other black (hot lead); a three-wire cable will have a white (neutral), a black, and a red (both hot leads) wire. In any cable with a ground wire, the ground may be either bare, green, or green with yellow stripes. Cable with up to three wires (including the ground wire) can be purchased in most convenient sizes for your work. However if you need four or more wires, you will have to run separate conductors yourself in conduit.

Two basic types of nonmetallic, sheathed cable, called type NM and NMC, are available. Type NM, commonly called *Romex,* can be used only in dry locations. It consists of several individual wires covered with insulation. Each wire is also spiral wrapped with paper and the entire bundle of wires is sheathed in an outer plastic cover. In type NMC, the individual insulated and bare ground conductor are all embedded in plastic. This type of cable can be used in all wiring applications, wet or dry.

The BX or armored cable mentioned earlier consists of a spiral-wound, overlapping and interlocking steel cover, with the individual insulated wires protected inside. A layer of paper is used around the wires and inside the metal sheathing to protect the insulation from abrasion with the steel jacket. Also, a bare ground wire is generally wound around the paper, just under the metal cover. Contact between the ground wire and metal cover grounds the metal sheathing and makes this still another path back to ground. The sheath is not a ground that you can depend on to exclusion of true bonding wire, but as a secondary path it can provide added protection. BX or other armored cable may not be used outdoors or underground due to the natural danger of corrosion and rusting of the metal covering and subsequent deterioration of the interior wiring.

Other Types of Wiring

Flexible cords and extension cords, such as those used on lamps and portable appliances, are normally made of small-diameter stranded wire imbedded in solid rubber insulation. These are called SP wires; for plastic insulation, the designation is SPT.

Wires designated for low voltage systems such as thermostats and doorbells

*Though it is a trade name, BX is a term that is widely used. We will use BX whenever we mean an armored cable.with a spiral metallic sheath wrapping.

have very thin conductors that are covered with a light layer of insulation. These wires may only be used in circuits operating at less than 30 volts and at low current levels.

From the discussions so far, you should be able to see the choice of wires you will need for a particular job. Cables with plastic-insulated copper conductors will probably be your best choice. Then, decide on the wire size, depending on the ampacity you want in a circuit (usually 15 or 20 amperes). Use a larger size if the load is a distance from the service entrance and voltage drop is significant. Finally, choose the type of cable you want to work with, metallic or nonmetallic, giving thought to the area in which it will be installed (wet, dry, above-ground, underground, etc.) and what the code requires you to use for a safe installation.

CONDUIT

When local codes require the use of conduit rather than cable or when more than three wires must be run between two points, you will have to consider using tubing called conduit (see Fig. 4-4) to protect the wires. Conduit is available in three types; rigid thick-walled, which requires external threading, like water pipe; a thin-walled, easier to use and cheaper type called EMT; and a flexible type called Greenfield. EMT comes in 10-foot lengths and in sizes up to 4 inches in diameter. It fits together and fastens to junctions boxes with various types of joint

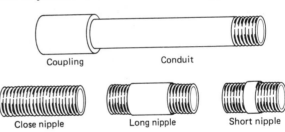

Coupling   Conduit

Close nipple   Long nipple   Short nipple

Rigid steel conduit with coupling and nipples.

(a)

Fig. 4-4(a). Rigid steel conduit is available in aluminum, steel galvanized and black enameled forms. Sizes range from ½″ to 6″ internal diameter and lengths to 10 feet. It is normally furnished with both ends threaded and a coupling on one end.

Fig. 4-4(b). Thin wall conduit, the most popular type for home wiring, is light and easy to bend and handle. The walls of this type of conduit are so thin that they cannot be threaded. Therefore, special couplings (see Fig. 4-5) sized to the conduit must be purchased to fasten conduit ends to junction boxes or to fasten lengths of conduit to each other.

Fig. 4-4(c). Similar to BX armor in construction, "Greenfield" flexible conduit is useful for runs involving many bends. It is available with internal diameters from ¹⁄₁₆″ to 3″ and coiled in lengths to 250 feet.

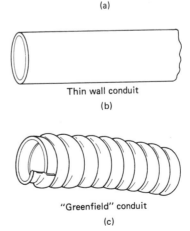

Thin wall conduit

(b)

"Greenfield" conduit

(c)

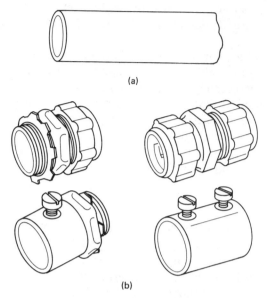

(a)

(b)

Fig. 4-5. Typical EMT thin wall conduit fittings.

connectors and special threadless compression-type fittings. Figure 4-5 shows some of the fittings that can be used with EMT conduit. For residential work, the most often used sizes are ½ and ¾ inches in diameter.

Once in place, conduit-run wiring is by far the best. However, it does take extra work to bend and shape even the thin-walled conduit around corners, fasten it together at the ends and fasten it to boxes. Also, the individual insulated wires of your system must be fished through each piece of conduit as you proceed. If you do use conduit, ask for THHN or THWN wires, which are thinner than most, or use wires that are prewaxed for easier pulling.

SURFACE RACEWAY

Surface raceway systems (Fig. 4-6) comprise metal housings that are designed for installation right on wall and ceiling surfaces. They are somewhat like conduit because they protect the wires with a metal sheath, yet differ from conduit because they are generally made with a cover that can be removed easily. This, plus the fact that they mount directly on the surface and are not hidden in the walls like cables or conduit, makes for easy accessibility to the wires. The initial installation is therefore simpler and changes can be made more easily. Raceways are often available in different finishes to match the wall or ceiling surface. Despite these advantages, you do lose something in appearance. Also, surface raceways are fairly expensive to purchase, and they are not always allowed by local codes.

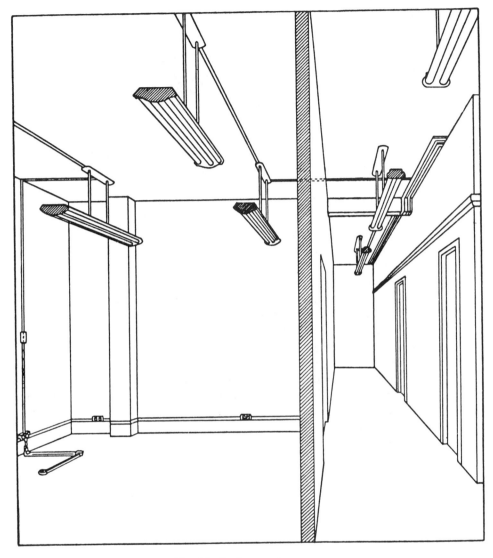

Fig. 4-6. Surface metal raceways.

SERVICE ENTRANCE

The service entrance panel contains the electrical system's main disconnect switch and the distribution network that separates the single incoming electrical supply into a number of branch circuits. The main supply and each of the branch circuits contain overcurrent protection devices that may be either fuses or circuit breakers as described in Chapter 3.

In both cases the input power from the electric meter comes through a single heavy cable at the top of the panel and enters the main disconnect/overcurrent protection element. In the fuse panel, the main overcurrent protectors are behind the left-hand fuse block. The two main fuses (cartridge type) are mounted behind the block. When the entire system draws more power than the rating of these fuses (but not enough through any one branch to take out its fuse or breaker), they will blow (open the circuit) to protect the system. Moreover, to disconnect all power you need only remove this single fuse block and the system is off. The second fuse block is to the right. In this panel, the right-hand fuse block contains two additional cartridge fuses of lower rating than the main fuse. These protect only a single 230-volt branch circuit that supplies a major appliance, most often an electric kitchen range or heater circuit. Note, also, that many fuse panels contain a switch lever that allows a more rapid shutoff of current than is available with a pull-out fuse block panel. Below the main fuse blocks are plug fuses for four individual branch circuits.

A circuit breaker panel costs more than a fuse type panel of equal capacity, but it has several advantages that make it a sensible choice, so sensible, in fact, that it is becoming the more common type. If the fuse panel is being rewired in an older house, it is advisable to replace it with a circuit breaker panel instead.

As mentioned in Chapter 2, circuit breakers are resettable circuit interrupters. When they trip due to an overload or a short circuit, they can be reset by throwing a lever or pushing a button, and there is no need to search for a replacement fuse. In the circuit breaker panel, the main 200-ampere circuit breaker protects the entire system and also acts as the main disconnect switch. The main breaker feeds power to the lower amperage breakers serving the branch lines. A typical 200-ampere panel may have room for up to twenty-four 115-volt circuit breakers or twelve 230-volt breakers (because each of these occupies twice the space of a 115-volt breaker). Note that the total current carrying capacity of the branch breakers may add up to more than 200 amperes, but it is highly unlikely that all circuits will be in use at the same time and loaded to their maximum capacity. In this panel, each branch breaker will trip if its rating is exceeded. The main breaker will trip only if the sum of all branch currents exceeds its rating.

To assemble these panels, start with two 230-volt and six 115-volt breakers, to service the range and air conditioner circuits plus six general purpose lines, then add 115- or 230-volt circuits at will until the total panel capacity is reached. Adding circuit breakers and additional branch circuits to this type of panel is quite simple—generally simpler than with a fuse panel—which is another advantage of the circuit breakers. The wiring of both types of panel is covered in Chapter 6.

Fuses

Most fuses in use today are of the plug type. They have a screw-socket base identical in thread and size to a light bulb base. Plug fuses are available in current ratings of 3, 6, 10, 12, 15, 20, 25 and 30 amperes. The most common sizes are 15,

20, 25 and 30 amperes. A fuse's current rating is clearly marked on its face so that it is visible when the fuse is installed in its socket.

Fuses present one major problem. If a circuit is designed for 15 amperes, for example, and you plug in so many lamps and appliances that the fuses blow quite often, you might be tempted to solve your fuse problem by installing a 20-ampere fuse. This solution may prevent the fuse from blowing, but at the same time it can create a fire hazard by permitting a higher current to pass through the fuse than the branch wiring can safely carry.

To eliminate this dangerous practice, as well as accidental up-fusing, we and the NEC recommend that type S fuses (see Fig. 4-7) should always be used in place of plug fuses. Type S fuses are made with three different-size screw bases that fit into special adaptors, which are easily installed in the sockets that former-ly accommodated any plug fuse base. Once adaptors are installed, the fuse will only fit into its proper socket. These fuses are made in three different current ranges, 1–15, 16–20, and 21–30 amperes. Within these groups, fuses of different ratings may fit the same socket, but these fuses at least limit the danger of up-fus-ing.

All type S fuses, as well as certain special plug fuses, provide a *time delay fea-ture*. The fusible link of these units only melts immediately if there is a true short circuit. In case of a simple, momentary overload, the strip will soften, but not melt, unless the overload continues for a period of time. This feature helps to pre-vent "nuisance" blown fuses in case of a momentary surge, which might occur, for example, when a high current-drain motor is first turned on.

Another type of fuse, a *cartridge fuse* (see Fig. 4-8) is generally used for two-pole (230-V) circuits, including protection for the service entrance conductors. These fuses are generally of the time-delay variety. They have a tubular, noncon-ducting body containing the fusible element, and metal end caps that snap into spring clip contacts in the fuse panel. Cartridge fuses with ratings greater than 60 amperes have metal blades protruding from each end. These snap into special

Fig. 4-7. A special screw-in insert is re-quired to adopt the fuseholder for type S fuses. (Graf–Whalen photo.)

Fig. 4-8. Cartridge fuses have no visible indicator to show if fusible element is still intact. (Graf–Whalen photo.)

spring clips and are most often used in the main disconnect fuse block of the service entrance panel.

Cartridge fuses are available in a wide current range up to several hundred amperes. They do, however, have one drawback: There is no visible external evidence when they have blown. If a cartridge fuse is suspected of being blown, it must be removed and tested either by substitution of another fuse or by an electrical continuity tester (see Chapter 8 for testing procedures).

Circuit Breakers

Circuit breakers are of a more complex design than fuses and are therefore more expensive. But they work more efficiently, are resettable many times, and they operate as switches as well as overcurrent protectors. For all this convenience, you should expect to pay a little more and be happy that they are available. Once installed, they seldom need replacement.

Breakers come in many ratings and styles. All are equipped with a front-mounted, switch-like lever that is manually operated to reset the breaker when it trips, or to shut off the circuit at will. The most common breakers used for residential work are either called single-pole, single-pole double (or piggyback), and double-pole. All are designed to snap into place in a service entrance panel designed to accommodate circuit breakers.

Single-pole breakers are rated at 15, 20, 25 and 30 amperes. They are used on 115-volt branch circuits that have been designed for this rated current and occupy one space of a service entrance panel. Twenty-four of these breakers protecting 24 branch circuits fit neatly in the 200-ampere service entrance panel described earlier.

Double-pole breakers are needed for handling 230-ampere service to large appliances, such as an electric range or central air conditioner. These breakers occupy two spaces of a circuit breaker entrance panel and are available in ratings from

20 to 200 amperes for residential use. Double-pole breakers can generally be recognized by their physical size and by the fact that they have two switch levers permanently linked together by a bar.

Piggyback breakers are designed to increase the branch circuit capacity of a panel when a number of lightly loaded circuits are desired. These units allow two branch circuits to be protected and controlled from one circuit breaker space and are available only in 15- and 20-ampere sizes. With these units, a service entrance panel with 24 breaker positions can be made to handle 30 or 40 lightly loaded branch circuits. Typically, a panel will contain 3 or 4 double-pole units, 10 or so single-pole units, 4–6 single-pole doubles, and still have room for growth through addition of more breakers and circuits.

Ground-Fault Circuit Interrupters

Ground-fault circuit interrupters, which were described in detail in Chapter 2, are available in several configurations. Some GFCIs are designed to replace ordinary circuit breakers in the service entrance panel. These, therefore, protect an entire two-wire circuit from any current imbalance between hot and neutral wires while also acting as a conventional circuit breaker. These GFCIs must have the same ampere rating as the circuit breaker they replace. This type is more sensitive than a circuit breaker and may cause some nuisance tripping as a result of temporary, nondangerous line surges. However, a GFCI breaker is still better than a common circuit breaker, because it provides the added personal protection against shock currents high enough to be lethal but not high enough to trip an ordinary circuit breaker. For bathrooms, kitchens, swimming pools, and outdoor circuits, GFCI breakers are the easy way to meet code and real safety needs.

Be sure to look for a GFCI unit with a Class A rating. This class of GFCI will shut off current within $\frac{1}{40}$ of a second after the onset of a leak as small as 5 milliamperes (0.005 amperes). Class B units will allow 20 milliamperes flow before cutoff; good enough to save your life, but why settle for less than the best when your life or that of a family member may be at stake?

Make sure you install GFCIs into all bathroom, outdoor, and garage receptacles and swimming pool circuits; they are required by the NEC. Also, for extra safety, they are certainly worth the investment for workshop, garage, or basement circuits, because these are particularly hazardous areas. GFCIs may also be required by local codes for kitchens, bathrooms, and other areas. To be sure, check your local code carefully on this point.

BOXES—THEY'RE A KNOCKOUT

There are, quite literally and bewilderingly, hundreds of kinds and sizes of boxes for use in electrical work. One thing they have in common is that the NEC requires that they be used wherever wires are spliced, connected together, or connected to terminals of any type of electrical equipment. This makes it necessary that you recognize the kinds of boxes you are likely to need in any wiring job.

Boxes can be of metal or plastic, but before using plastic check your local codes to see if and where they can be used. Presently offered boxes are octagonal, rectangular, or square in shape, varying from 1 ½ to 3 ½ inches in depth. They are usually provided with knockout holes at different locations. These holes are somewhat like perforated paper. Sections of the box walls are weakened by being partially cut through; you can therefore easily pry out or knock out the weakened section at the location you want to have a wire enter or exit the box. Then, with a smooth hole in the box, you can insert the end of a cable and fasten the cable outer cover to the box with a cable clamp. The wire ends of the cable thus extend into the box and can be stripped, shaped, fastened together, or manipulated as you wish. More important, the actual wire connections will be isolated from flammable building material if anything goes wrong when current is on. That is the real purpose of boxes: keeping connections enclosed in a dead space that limits their physical contact with materials that can burn.

Figure 4-9 shows boxes that are representative of presently available types. Some are equipped with factory-installed brackets on the sides or back of the box. These can be used to nail the boxes in place to building studs or beams and save time in installation. Some boxes simply have drilled clearance holes for nails or screws and leave to your ingenuity the method by which they will be fastened at the locations you must have them. Some boxes are designed for covers, others have an open side that mounts flush with the outer wall surface so that an outlet receptacle, light fixture, or switch can be installed. Some boxes can be stacked on one another or separated into pieces and mated with other pieces for expansion. Boxes with covers are generally used as junction boxes for splitting circuits into several branches or as pull boxes. Pull boxes are used with conduit to make it easier to pull wires through long lengths of conduit. They may only be located where they can be permanently accessible without damaging the building structure (by code).

After you decide how particular boxes will be used and where you would like to mount them, your hardware supplier can probably help you decide which boxes to use. There is a limit to the number of wires that can legally be installed in a box of given size and this must also be given consideration in your planning.

Box Mounting Arrangements

When a box is positioned alongside a stud or beam, the simplest fastening method is to nail it directly to the wood. You can use boxes with attached brackets or insert several nails through one side of the box and out the other. Most boxes have drilled clearance holes to allow this. Nails are permissible but remember, they take up room inside the box and the wires must be tucked in around them. You would probably only use this for an uncrowded box with single switch and few wires to accommodate.

If you must mount a box between studs or ceiling joists, the box hanger shown

Octagon boxes for use with fixture or junction outlets. These are for use with conduit, or armored or non-metallic cable. Available in steel with half-inch knockouts or with cable clamps. Important hint: use bar hangers with studs wherever ceiling fixtures are to be installed. Recommended for in-home use.

Use steel boxes for switch, receptacle and bracket outlets in the home. Available with or without clamps. Gang two or more by removing one side plate on each box. Then hook boxes together.

Steel outlet or switch boxes with wall mounting bracket. Mount brackets to allow correct depth for lath (or sheetrock) plaster or paneling. Available for use with connectors or with handy built-in clamps. Warning: use only in new work.

At left: beveled corner box. Fits into wall opening in old buildings. Has clamps for loom or non-metallic sheathed cable. No connector needed. Side-bracket box (center) used with cover (right) in new work, wherever extra space is needed for wires. Used for switches, receptacles, bracket outlets. Combination ½ and ¾-inch knockouts, requires connectors. Use with armored, non-metallic or plastic cable, or with conduit.

Fig. 4-9. Various available boxes. Courtesy of Sears, Roebuck and Co.

in Figure 4-9 is the best choice. The length of the hanger is adjustable to match the joist spacing, and the box may be positioned anywhere along the hanger length. Fasten the box to the hanger by removing the center knockout of the box, inserting the hanger stud, and fastening with a locknut. Then, nail the hanger ends to the wood so that the front edge of the box will be flush with the wall or ceiling. Preassembled boxes with hangers are also available. They offer most of the adjustments just noted and are handy to use.

RECEPTACLES

The receptacle or outlet can be thought of as the working surface of your wiring system. The system wiring threads through the house, largely hidden from view inside walls and ceilings. Then, at planned intervals, boxes are mounted in the wall and receptacles are mounted on the front surface of the box so that they are flush with the surface. These receptacles are connected to the wiring in the box and accommodate the plugs of line cords from appliances or lamps to provide electric power.

The most widely used receptacle is a duplex (double) unit, with recesses to accept the blades of two appliance plugs. Normally, both halves of a duplex outlet are connected together so that when one is connected into a circuit, the other also becomes a source.

At this point, we should distinguish between old and new work. If you replace a receptacle on an old outlet box, you will often find that the outlet box is not grounded. In this case, you must, by code, use a two-wire duplex receptacle, with holes for only two blades. This will also be true if you are adding a receptacle to an existing ungrounded branch circuit. The reasoning is logical. If you added a three-hold receptacle and had no way of grounding the ground blade, someone could be misled into thinking the plug was grounded when it wasn't! To keep your installation legal, you must replace like with like. Otherwise, you must re-wire.

For any new work (either a new system or a new branch circuit), you must use receptacles of the three-wire, grounding type in accordance with code requirements. These receptacles can accept either two-blade plugs, which are still standard on lamps and small appliances, or the three-blade grounding plugs found on many tools and larger appliances. The assumption here is that the wiring leading to this new work is also in keeping with the code. The grounding receptacle can be recognized by the three-hole pattern of holes at each plug position. Every grounding receptacle also has a special terminal, colored green for identification, for connection of the ground wire from the power cable.

Receptacles equipped with screw terminals are wired by stripping the insulation from a short length of the wire end, forming the conductor into a loop, and fastening it to the terminal by means of the screw on the receptacle. However, you can also consider using receptacles with the more convenient push-type terminal. They might cost a bit more but are much easier to use. The stripped end of the wire is simply inserted into a hole in the rear of the receptacle housing. Bared wire ends are held fast automatically by grippers inside, and can only be released by inserting a screwdriver into a release hole nearby.

Other available receptacle types include covered receptacles, with accommodations for three plugs in the space that normally take only two, or outlets with hangers for wall clocks and space to store excess cord. There are also safety outlets with spring-loaded covers that twist shut so children cannot easily poke things into openings unoccupied by plugs. For outdoor work, there are weatherproof outlets, equipped with hinged or screwed-on covers. For special applications, such as powering two high-current appliances, there are split receptacles with breakaway tabs between the two duplex outlet halves so that each half can be wired from separate branch circuits.

SWITCHES

Switches take so many forms to meet so many specialized needs that we can describe only the types that you are most likely to use in a residential system. For such use they are made in three models: single-pole; single-throw (SPST); and three-way and four-way. Like receptacles, most switches are mounted on the open side of electrical outlet boxes, so that they are flush with the wall surfaces, and are connected by screw or push terminals to the wires inside the box.

The most common type of switch is the single-pole, single-throw, the familiar household light switch with an ON-OFF operating lever. These switches are made with two terminals—one terminal is connected to the power source hot side, the other to the load. As such, the switch is in series with the source and the load. This switch can control a light, fixture, receptacle or group of receptacles from a fixed location and is installed so that when the switch lever is pointing up, the legend ON is visible and the contacts are closed to allow current flow. The switch may be mechanical, in which two contacts are pushed together or forced apart, or it may be a mercury-type in which a small pool of mercury bridges two contacts when the switch lever is moved to on.

A three-way switch is used to control a light or series of lights from two different locations, such as at the top and bottom of a staircase. This type of switch has three terminals and the lever is not marked ON and OFF. In this switch type, there are two contact positions for the moveable switch arm. It is called a single-pole, double-throw (SPDT) switch.

Four-way switches have four terminals, are not marked ON and OFF and are used to control the same outlets as the single or three-way switches. These, however, allow control from three locations. In practice, you use two three-way switches, plus one four-way switch. The four-way switch has *two* contact positions for each of two separate moveable switch arms. It is called a double-pole, double-throw (DPDT) switch.

When you decide which type of switch you need, check that the selected switch has the current capacity for the job. Most UL switches are rated at 10 amperes, 125 volts, or 5 amperes, 250 volts and these will handle most household service. If you have a heftier job to do, check with your supplier for a heavy-duty switch that is rated for the larger current you need.

Special switches such as the following can also be used to improve your system:

Pilot-light switches that tell when a remotely located motor or light is on.

Switches with neon-lamp lighted lever so that the lever can be found easily in the dark when the switch is off.

Delayed toggle switches that have a turn-off feature giving several minutes of power ON after the lever has been moved to the OFF position.

A timer switch that provides the same delayed shutoff but for longer periods and that may have a built-in clockwork mechanism.

A Special Word about Dimmer Switches

Thanks to modern electronics, a new kind of switch has become available that allows control of incandescent lamp brightness by turning a knob or sliding a lever. Technically such a device is known as a *dimmer switch,* but actually it is a combination of a switch and an electronic circuit that can gate controllable amounts of power to a lamp, depending upon the setting of a knob or lever. The switch portion of a dimmer is the same as the common SPST wall switch. But the electronics consist of a solid-state device known as a *triac* that is in series with the power source and the lamp load.

A triac is a device that can electronically switch from a nonconducting to a conducting state in just thousandths of a second. This means that it can chop the available power in the alternating current wave into any desired duration of ON time each time the a-c line goes through a cycle. In practice, this means that the triac can be made to switch on, then turn off in step with the polarity changes of the voltage of the a-c line. It allows specific amounts of power to be pulsed through a lamp for any given setting of a control knob or lever. Essentially, the amount of power that reaches the lamp is proportioned to the setting of the control knob or lever, and so the lamp responds by glowing at a brightness level that is proportional to the average voltage and current flowing through it. The result? The lamp can be adjusted to glow from full brightness (power on for the complete a-c cycle) to off through any desired increment of brightness.

Lamp dimmers are popular items for mood lighting and find ready replacement use as an alternative to ordinary wall switches that control incandescent lamps. However, they cannot be used for control of fluorescent lamps. They should also not be used to control combined circuits that feed wall outlets in addition to incandescent lamps because common motorized appliances (e.g., fans, air conditioners, kitchen appliances, etc.) do not respond to the pulses of current supplied by dimmers. So don't expect more than being able to control the brightness of room lights.

If you decide to use a lamp dimmer, add the wattage rating of the lamps you want to control in the circuit and buy a dimmer that meets this rating. A lamp dimmer fits neatly into the same box space as a common switch. You may find that the dimmer has wire leads that must be connected to circuit wiring through solderless connectors. Likewise, don't expect to be able to wire a dimmer into an upstairs/downstairs three-way switch circuit without some compromise. At best you will achieve a single location dimming function overriden by a switch at a different location.

Used within its limitations, a dimmer switch is a handy device. But it is best to think of it as a specialty device for use on circuits involving only lamps you want to increase or decrease in brightness.

LIGHTING FIXTURES

Lighting fixtures can be used simply for good illumination or for special decorative effects, both indoors or out. Illumination for reading or close work can generally be provided by a single lamp plugged into the nearest wall receptacle. Lamp arrangement is then mostly a matter of convenience and your personal taste. Lighting fixtures, though, will more likely involve area lighting from a ceiling fixture or indirect lighting that throws a background of diffused light over the walls. Whatever the use or type selected, lighting fixtures have two things in common:

They must be supported mechanically so that they will not fall and there must be a provision made to properly connect the fixture wires to the house wiring.

A common ceiling fixture mounts to an outlet box fastened to a hanger between two ceiling joists. The fixture may be fastened to the box by a variety of methods, such as metal straps or threaded studs that fasten to the top of the fixture. (See Figure 4-10.) In either case, wires from the fixture are connected to the wiring system inside the box.

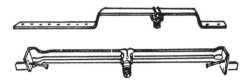

Fig. 4-10. Bar hangers.

TOOLS

When you design your system and buy all the component parts you are still not quite ready. The parts go together easily enough, but they must be cut to length, bent to shape, and fastened together. The walls and floors must be drilled and cut, plastered back, and so on. All of these tasks require the proper tools. The first suggestion we have is to buy the best you can. Anything you do yourself will save much more than the cost of the tools, and the better your tools are, the easier your job will be. Our second suggestion is that you rent or borrow some special tools that you know you will only need temporarily.

A good basic complement to the basic tools you probably have already is shown in Fig. 4-11(a). Assuming you are a householder, we will not discuss the

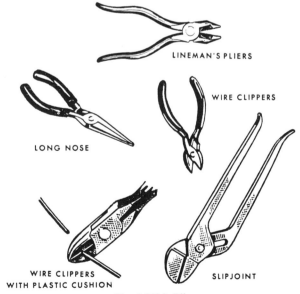

Fig. 4-11(a). Pliers.

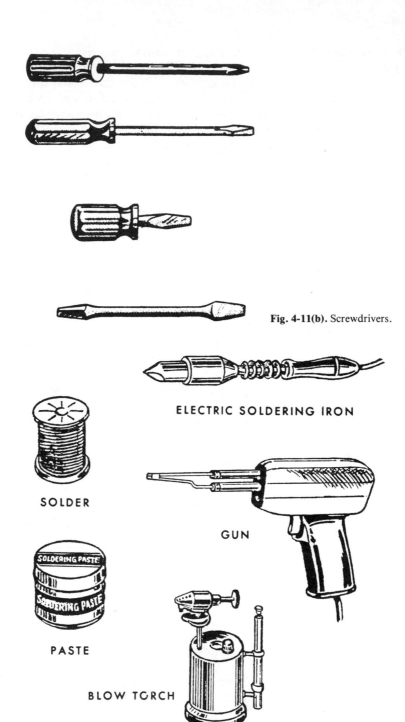

Fig. 4-11(b). Screwdrivers.

ELECTRIC SOLDERING IRON

SOLDER

GUN

PASTE

BLOW TORCH

Fig. 4-12. Equipment for soldering.

use of most of them. The pliers, however, could stand a quick review. Lower right is a slip joint plier, which you may already have, but is yours equipped with insulated handles? We know you will *never, never* work on a circuit that is hot, but buy a new plier with insulated handles to be absolutely safe. Note that there are several other types of plier in the picture. Also shown is a long-nose or needle-nose plier used to shape end-loops in a conductor for attachment to a screw terminal. The diagonal cutting plier, also called wire clippers, is used to cut wire to length.

Several types of screwdrivers are shown in Figure 4-11(b). When you get down to the work you will find the need for several of different sizes and blade types. An automatic-return screwdriver will easily pay for itself in saving your wrist muscles on a job. Also, an electrician's screwdriver will be useful. It is made with a long slender blade the same width as the shaft to get at screws that are deeply recessed.

The drilling equipment shown will probably cover most of your needs. The electric drill is handy for general drilling, such as pilot holes for installing screws. The brace and bit should be used for drilling large holes through beams or studs or between floors in which there are several layers of wood. The saber saw is invaluable for cutting sections out of a plaster wall or ceiling for installation of outlet boxes of various sorts.

Figure 4-12 shows some equipment you will need for soldering. Other than for repair of small appliances or lamps, you may not find a need for soldering equipment, so wait until the need arises before you purchase any of these items. Generally speaking, we recommend the soldering gun over the soldering iron. Solder has been used less and less in the past few years for electrical installation work.

Figure 4-13 covers use of a pipe hickey which you will need if you are working with conduit. The long pipe with the curved bracket on the end is used to bend conduit for installation around corners. Another tool, known as a conduit bender is also available, and makes smoother bends. The conduit can easily be cut with a hacksaw, but you will need a reamer to deburr the inside of the conduit ends after cutting. Any burrs that are left could be particularly damaging and dangerous to the insulation as you pull the wires through the conduit. Another tool you will need is *coiled fish tape*. It has a hook at one end and is used to "fish" for cable that is to be pulled through the conduit. The stiff tape is first pushed through the conduit, then the wires for that run are attached to the hook and fed into the conduit as the tape is pulled out the other end. Incidentally, fish tape is useful for re-wiring jobs. It can be used in many ways to snake cables through walls, floors, and ceilings, as described in Chapter 5.

Although the tools discussed here will help with the electrical portion of the job, you will probably need additional tools for carpentry, masonry and plaster work, digging, and climbing. These tools are the sort that can be acquired as needed.

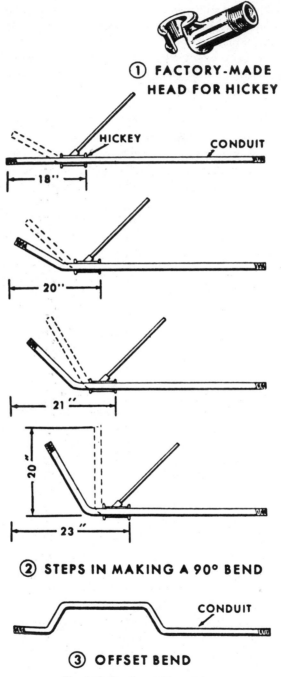

① FACTORY-MADE
HEAD FOR HICKEY

HICKEY   CONDUIT

|← 18" →|

|← 20" →|

|← 21" →|

20"

|← 23" →|

② STEPS IN MAKING A 90° BEND

CONDUIT

③ OFFSET BEND

Fig. 4-13. Bending rigid conduit.

5

METHODS

Wire Terminations–Wiring Routes–Connection Boxes

The word "methods" covers a multitude of projects for a wiring installation. We will use it to mean any general technique of manipulating wires or other parts to integrate them into a working system. This includes clean up and patch up, but will not cover the electrical circuitry make-up, which is discussed in the remaining chapters of this book. We could say that this chapter details the *mechanical functions* in the installation of an electrical wiring system. Though the wires and cables are not the first things that are installed, we will discuss them first because they are so fundamental to every job. Master these techniques well and all other steps will be more understandable.

WIRE CONNECTIONS

To make your system function, wires must be connected end to end to one another and to receptacles, switches, and other circuit elements. This is probably the single most important aspect of the whole job. All such connections must be made carefully and inside junction boxes. If you do not follow established procedure, the system will either not perform, function intermittently with many breakdowns, or, worst of all, become a shock or fire hazard.

First, as you know, most conductors are covered by insulation. Some of it must be stripped away at each end of the wire to expose bare conductor material to make a connection that will transmit electricity from one wire to another or from a wire to a circuit element. For cabled wires this means two layers of covering, the outer armored or nonmetallic covering, and the inner insulation around each wire. Figure 5-1 shows how to remove the outer nonmetallic

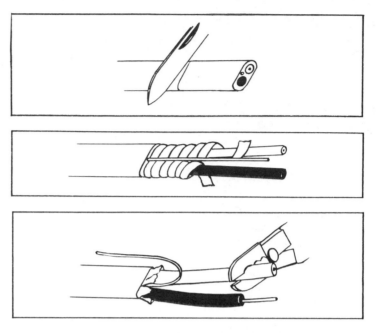

Fig. 5-1. Preparing Romex (NM) cable. (a) Carefully cut outside fabric sheathing 6″ to 8″ from end and strip away from inside insulated conductors. Avoid cutting into insulation of conductors on inside. (b) Remove outer wrapping from the insulated conductors and pull back firmly against the outer jacket. Carefully cut the wrapping flush with the jacket. (c) Strip the insulation on each of the conductors one inch from the end, so that the copper wire is visible.

covering from Romex or similar cable, while Figure 5-2 illustrates how to handle armored (BX) cabling. For BX, note that after cutting and stripping the spiral-wrapped armor from about the last 8 in. of the wires, fiber bushings must be inserted under the armor to prevent the raw ends of armor sheathing from cutting through the insulation of the wires.

A knife is a good tool for stripping insulation from the last inch or so of individual wires. Hold the knife at about 60° to the insulation, not at right angles. This will help to prevent nicking the conductor, which could weaken it. *This is particularly important for thin stranded conductors.* After you have cut completely around the insulation, pull the insulation off the end, leaving a short length of bare conductor projecting. Clean off all traces of the insulation from the bared end.

Figure 5-3 shows a wire stripper in use. The wire stripper is a handier tool than a knife if you are careful to use the right tool slot. These tools are made for different types of conductors and have holes that fit various sizes of insulated wires. With the wire in the proper hole, the handle is squeezed to cut through the insulation. However, the cutter jaws are arranged so that they cannot close

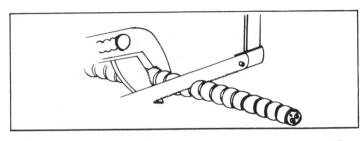

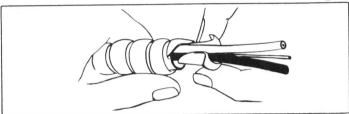

Fig. 5-2. Preparing BX cable. (a) Hold armor securely and use hacksaw at slight angle to cut through armor approximately 6″ from cable end. Be careful not to cut into the insulation inside conductors. (b) Twist armor loose and discard. Strip one inch of insulation away from conductors. Insert fiber insulator around conductors and push inside the armor against the cut.

beyond a certain diameter, preventing nicking or damaging the conductor itself— *if you use the right size slot.*

Now that you have a bare conductor end, you are ready to connect it to the next electrical element of the wiring system. If the next element is another wire, the wires may be connected by splicing (see Fig. 5-4) and soldering (see Fig. 5-5); or by means of mechanical solderless connectors such as wire nuts and split bolts (see Figs. 5-6 and 5-7). Solderless connectors (wire nuts) are only permissible where there is no strain on the wire, such as in an outlet box. Solderless connectors are becoming more and more popular, while soldering has decreased in use recently. This can be explained by the fact that wire nuts are simpler to use but still make a strong bond. The wire nut presses the conductors together mechanically and produces a good electrical connection as well. What does the code say about joints between two wires? Well, the code has very simple instructions, as follows:

1. Mechanically, the joint must be as strong as a continuous piece of wire. (As mentioned, an exception is made to allow wire nuts where there will be no strain on the wires.)

2. Electrically, conduction from one piece to the other must be as good as a single piece of wire.

3. Insulation covering the joint must be as good as the insulation on the individual wire.

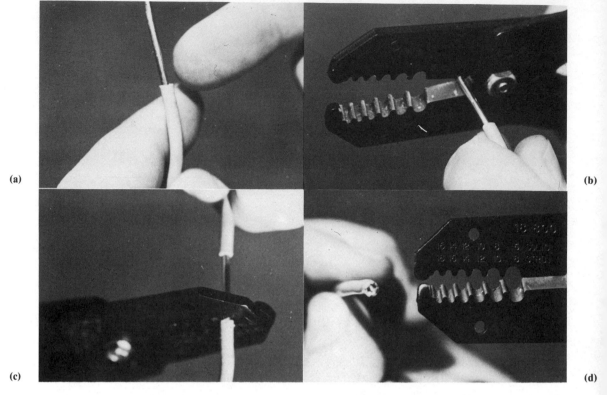

Fig. 5-3. Stripping insulation. (a) The right way to strip insulation with a knife. Cut through the covering and completely around the conductor at a shallow angle to reduce danger of cutting into (and weakening) the conductor underneath. (b), (c), and (d) The wire stripping tool is still more handy for removing insulation. Noches of different sizes, which fit standard gages of wire are cut into the jaws between the handles. Close the handle firmly on the wire and the insulation is cut clear through; the insulation may then be slipped off the wire end. (Graf–Whalen photo.)

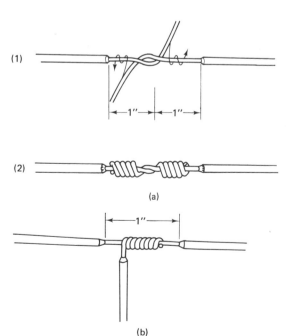

Fig. 5-4(a). The Western Union splice is a good connection. With three inches of bared conductor on each wire, start the splice by making an L-shaped bend in each wire as at the left above. Then twist one wire in 6-8 turns forward around the second wire. Do the same in reverse for the second wire. Note that wrapping forward creates many points of contact between each wire. Doubling a wire back on itself would not make a good connection at all. **(b).** The tap splice is used to connect a wire at an intermediate point of a continuous wire. Strip an inch or so of insulation from one wire and twist the bare end of the tapping wire lightly around the bared conductor. As with all these splices, solder the connection and insulate with wrapped layers of tape.

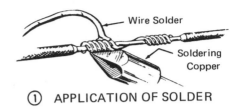

Fig. 5-5. Application of solder to a Western Union splice. Before soldering check that solder iron (or gun) tip is clean and shiny. Hold the solder tip to the wire until the wire becomes hot enough to melt the solder. In the cross section (lower left) the solder has not penetrated completely into the coil since the wire was not hot enough to melt the solder well.

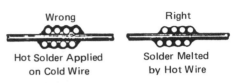

② RIGHT AND WRONG SOLDER JOINT

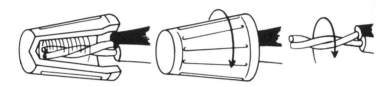

Fig. 5-6. Solderless connectors contain a tapered metal spring inside an insulated cap. To use, twist bared conductor ends together in a clockwise direction, insert into the connector and twist the connector clockwise to fasten firmly onto the wires. Solderless connectors are available in many sizes for holding different sizes or numbers of conductors. Courtesy of General Switch Company, Middletown, N.Y.

Fig. 5-7. Split bolt connector. Courtesy of Sears, Roebuck and Co.

Wire splices must be made with care to meet these criteria. Three general types are used: the pigtail splice (not illustrated because it is identical to the wire joint under a wire nut), the Western Union splice (Fig. 5-4a) and tap splice (Fig. 5-4b). Codes require that these splices must be mechanically and electrically secure before soldering. Also, they must be soldered and they must be covered with layered windings of insulating tape.

TERMINAL CONNECTIONS

Connecting a wire to a screw terminal of a fixture switch or receptacle is generally accomplished by forming an open loop at the very end of the bare conductor and clamping this under the screw head as shown in Figure 5-8. Needle-nose pliers are the best tool for this procedure. Be sure that the bared conductor is not too long. Insulation should come close to the terminal screw, leaving just enough room so that the insulation is clear of the screw head. Also, the conductor should be positioned under the screw head so that as the screw is tightened on the loop, the screw tends to close rather than open the loop. Figure 5-8 illustrates this technique. Most commercially available receptacles are equipped with captive screws that can back out of the housing only part way so they do not fall out and get lost.

Before leaving this subject, we should mention one other, still more convenient type of connection. In this receptacle, the straight, bare conductor is inserted into a hole in the rear of the receptacle housing. Serrated jaws in the hole grip the conductor, making a firm mechanical and electrical joint. The grip is so firm that the only way these conductors can be removed is by insertion of a thin blade screwdriver into a release hole alongside the conductor hole. These receptacles are often made with a molded-in gauge that shows just how much conductor should be bared for effective connection.

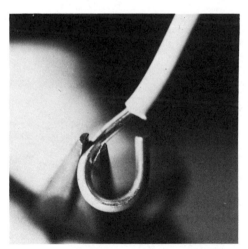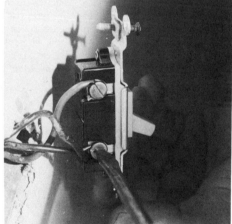

Fig. 5-8(a). To connect a wire to a screw terminal, remove about ⅜″ of insulation and form the end of the conductor into an open loop using needle nose pliers as shown. You can close the loop after you have hooked the loop over the shank of the receptacle screw. (b) Most receptacle screws are captive type so that they cannot be removed and lost. Back the screw out far enough to clip the loop over the screw shank just under the head. Use care to hook the loop to the screw with the opening on the right, and tighten the screw firmly down on the loop. With the loop fixed as shown the screw tightening will tend to close the loop rather than to open it. (Graf–Whalen photos.)

RUNNING CABLE AND WIRING

Your house can be considered as a huge skeleton to which you will add a threadlike network of nerve channels ending in outlets, fixtures, switches, and appliances. How the wires get from here to there is a study in itself. We will indicate some of the general routes and techniques; you can then improvise the details as required.

First, we have to distinguish between "old" and "new" work. If you add a new room, finish an attic, wire a bare garage structure or an entirely new house, the job is "new work" and is relatively simple because everything is so accessible. The skeleton is there before your eyes as construction proceeds. You can walk from room to room through the studs, nail or screw-fasten the boxes at any point you wish and run wires by the shortest route before the finish paneling and ceiling boards are added. Some of the more common methods of fastening outlets and ceiling fixtures if you have a bare wood structure with which to work include nailing the boxes directly to the joists using holes or mounting brackets on the boxes, and attaching ceiling fixtures to hangers nailed between two ceiling beams as shown in Fig. 4-6. Cables between the boxes and fixtures must be supported in any run; fasten BX with heavy staples and nonmetallic cable with metal straps or staples. Fasten them within a foot of every box and at 3-ft. minimum intervals on exposed beams or studs. Better yet, drill through the wood and thread the cable through holes. When conduit is used, it can be fastened and bent to shape and the wire fed through to the boxes before the walls are completed.

"Old work", though, is not quite as easy. More often than not, just running a cable is a task in itself that may involve two people. You have to deal with finished wallboard or plaster walls edged with molding, finished floors, and ceiling. Naturally you don't want to leave unsightly gashes or gaps in the plaster or ceiling and the catch phrase must now be: *Proceed slowly and steadily.* BX, plastic cable, and nonmetallic sheathed cable can all be used because they are flexible and can be pulled through the walls and floors easily. The NEC prohibits the use of flexible extension cords inside walls in place of cable. Whichever cable you use, you will probably find that old work takes more material, because you have to find the most accessible route that will avoid tearing up walls and floors. In old work you need only strap-support the cable in exposed portions of the run.

Do you have to run your wiring clear around the perimeter of a room? Try removing the baseboard molding along the route. You know that molding covers the carpenter's sins. Maybe under the baseboard and around the door molding there is a gap over half an inch wide into which you can stuff a cable. If it is not there, you might consider chiseling out a channel in the wallboard or plaster that will be covered later when you replace the molding. Fasten the cable with staples or straps and run it at will to your new outlet. If the wallboard or plaster is not thick enough, chisel a notch in the front of each stud you have to pass. Be careful when reinstalling the molding so that you do not nail into the cable or squeeze it with the baseboard.

If your service entrance is in an unfinished basement, you will naturally run the new cabling along the exposed basement ceiling beams until it is under the wall section in which you want an outlet. If the cable must run at an angle to the beams, nail a 1 x 3 board to the beam bottoms and staple (BX) or strap (non-metallic) cable to the board. For along-the-beam routes, cables may be stapled or strapped to the side of the beam. The cable must be supported at least every 4½ ft (overhead in the basement make this closer). It must also be supported within a foot of every box. Where the cable bends around a corner, the turn radius should not be less than 10 times the cable diameter. Drill from the basement into the upstairs wall core as shown in Figure 5-9. A long extension bit for a brace is most helpful for this job, which entails drilling at angles through several layers of flooring, floorplates, studs, and so on. This same technique can be used effectively when running cable through an unfinished attic. Drill down at carefully measured positions to gain access to the wall core just between the studs where you want to install an outlet.

Now that you have a hole from basement to wall core, you want to run the wire. Luckily, pushing cable is not exactly like pushing a chain; there is some stiffness to the cable and that will help. However, it is almost always easier to pull from the other end. At this point, the technique of connecting the wires using fish tape becomes valuable. Fish tape is a relatively stiff metal strip with a hook on the end. If one man works from the basement and pushes the tape through the drilled hole, a second man upstairs, peering through a formed outlet hole in the plaster, can catch the tape and attach the cable to be pulled back through to the basement. You need to measure carefully to be sure that your drilled hole enters the same space between wall studs that the outlet hole occupies.

Other fish tape applications are shown in Figure 5-10. In a completely "blind" location, one hooked tape can be fed in from above or from an adjacent

Fig. 5-9. A brace with a long extension bit for drilling at angles through flooring. Courtesy of Sears, Roebuck and Co.

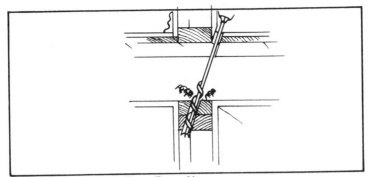

Pulling Wire Through Floor From Above

If pulling a wire from the room above, first remove the baseboard. (When the wire is drawn from the attic this is unnecessary.) Next locate the diagonal beams which cover the wall area you wish to fish the wires through. Drill a hole with an 18-inch bit and a brace.

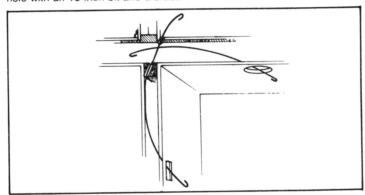

Feed one 12-foot fish tape through a hole in the ceiling where the fixture is to be placed, and run another 12-foot tape from the room above through the electrical box hole in the wall. Pull both tapes slowly taut until the two hooks catch.

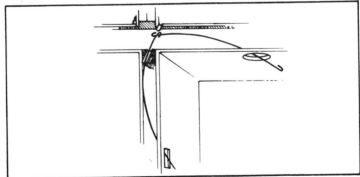

Secure the wires to the hook on the fish tape at the wall receptacle, and carefully pull the tape at the ceiling fixture hole until about eight inches of wire is exposed through the hole. Disconnect the fish tapes.

Fig. 5-10. Pulling a wire through from the floor above. Courtesy of General Switch Company, Middletown, N.Y.

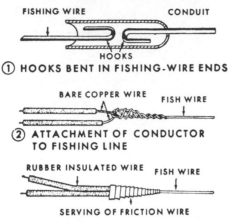

① **HOOKS BENT IN FISHING-WIRE ENDS**

② **ATTACHMENT OF CONDUCTOR TO FISHING LINE**

③ **ATTACHMENT TAPED OVER**   Fig. 5-11. Fish-wire pulling.

room while a second man feeds his hooked tape into the blind area. By twisting and poking you can tell by "feel" when contact is made; then, careful pulling will hook the tapes together so that one tape end may be pulled back through. Then, the cable may be attached (Fig. 5-11) and pulled through so that wiring may be completed. It sounds easy, and it is, but without the fish tape it would be nearly impossible.

Note that replacing old wiring is sometimes easier. Gain access to both ends of the wire and check that it is not fastened deep inside the wall. Then, hook new cabling to one end and pull it through to the opposite end as the old wire is pulled out. Splice the old wire to the new wire. Just be sure that the splice is tight and smoothly wrapped with insulating tape so it will not catch on any obstruction as it is pulled through. Next, pull out the old wiring and at the same time you will be pulling the new cable into place.

In some installations, you may want to use surface raceway wiring. Several UL approved forms are available and are approved by most local codes. Non-metallic types are simpler but metallic raceways provide the widest range of application. Surface wiring was the most common procedure in the early days of electricity. It is still easiest of all but, because it is not hidden within the walls, it is not the most attractive installation. It should be used in rooms such as basements, workshops, or garages where esthetics can give way to practicality. For old work, surface wiring minimizes the need to cut into walls and allows for later additions and expansion with very little work. Even though materials are more expension per running foot, they do save so much work that you might consider surface wiring in particularly tricky places where you cannot find a convenient route through the walls.

INSTALLING CONDUIT

As mentioned earlier, conduit is a seamless tubing of thin-wall, thick-wall, or flexible metal construction. If you must use it, your local code or building department can tell you what form it must take. If you are free of code constraints, you might still think of using thin-wall conduit as a premier installation that will do a beautiful job. Conduit is corrosion resistant inside and out and, furthermore, smooth-finished inside so that wires may be pulled through without damage to the insulation. Conduit comes in 10-ft lengths of various diameters. The diameter that you will use depends on the number and sizes of wires to be installed. A ½ in. (inside diameter) conduit will carry four No. 14 wires or three No. 12 wires; ¾ in. will carry four No. 10 wires, five No. 12 wires or three No. 8 wires, and so on.

Installation of wires in the conduit is accomplished by use of fish tape as described previously for old work. The only difference is the use of pull boxes at intervals of several conduit lengths to ease the process. Installation of the conduit itself is a different matter. To install conduit in an old house might require reconstruction of the entire structure. Wallboard and plaster walls, ceilings and floors would have to be torn to pieces to accommodate the tubing. For this reason, codes do not require use of conduit in old work and we will limit our discussion to new installations.

Thin-walled conduit is cut using a tubing cutter or by means of a fine (32 teeth/in.) hacksaw. Also it can be bent using a special tool called a pipe hickey. Conduit tools can be obtained from electrical distributors who supply professional electricians. Because wires are pulled into the conduit after it is installed, it is important that bending be accomplished carefully so that the interior diameter is not decreased. The bending tool available on the market can do a nice job if you make your bends smooth and gradual. For any type of conduit, the code specifies that the radius of the bend must be six to eight times the inside diameter of the conduit. Also, more than four-quarter (90°) turns or their equivalent (360°) are prohibited in any one run of conduit, one run being the conduit from one box to the next.

For new installations, conduit must be built-in by either drilling holes through studs, beams, and joists and inserting the conduit or by supporting the conduit at intervals (within 3 ft at least of every box, fitting, or outlet). If it must run along a series of studs, chisel out a notch in each stud and recess the conduit into the notches. Conduit must also be bent around corners or in slight jogs to align with the knockout hole of a box.

As mentioned earlier, pull boxes are often inserted into conduit to make pulling of wires easier. Use a fish tape to pull the wires and apply wire-pulling lubricant as you proceed. It is best to pull all wires at once, but don't do this step until wall plastering is completed. Make sure the wires you choose conform to standard installation color code (see Chapter 8). If a length of conduit has several intricate turns, you could use a pull box for each length. However, a straight run would only require a pull box for each two or three lengths of conduit.

Thick-walled conduit is threaded like plumbing pipes (½ to 1¼ in. inside dimension) and fitted together as well as to outlet boxes by means of threaded connections. The thin-walled conduit is, however, much easier to work and can be connected end to end, or to a box by slip-joint compression couplings. These couplings consist of a body plus a split ring whereby tremendous pressure is exerted on the outer shell of the conduit when the nut is turned down tight. This makes a mechanically secure connection with low electrical resistance, the conduit serving as bonding conductor back to ground when properly installed.

INSTALLING BOXES AND FIXTURES

Some typical locations for outlet boxes and switches are described in Chapter 3, namely, with the center of the outlet box about 12 in. above floor level or 8 in. above a counter top and a switch about 4 ft above the floor. These dimensions hold for old work or new; hanging the boxes in old work is your next major problem.

With new construction and open skeleton studwork, hanging boxes is a cinch. You just nail the box by means of an attached bracket or by inserting nails through one side of the box and out the other, then into the exposed stud. With old work though you have more of a problem. Is your wall wood lath and plaster? Refer to Figure 5-12 for installation. Cut through the plaster first at the approximate location and position the hole so that the center of the box will be directly over a full lath. Cut this lath clean through but only notch out the lath above and below. Then screw-fasten the box to the uncut portions of the top and bottom lath.

For plasterboard, be sure you are not astride a stud and draw the outline of the box on the plaster. Careful measurement or tapping the wall for a hollow sound will help, but a stud finder will also be worthwhile. (A stud finder is a device that magnetically senses the nails in the stud when it is passed along the surface of the wall.) Drill a starting hole in the corners and cut the box outline in the plaster with a key-hole saw. The most important technique to emphasize here is careful measurement. Is your wire going to drop down between two studs, from the attic, or up from the cellar? Be sure you cut the opening in the stud area where the wire will run.

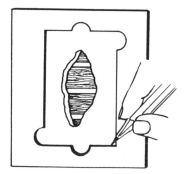

Fig. 5-12(a). To install a box on plaster wall, cut through the plaster to expose the lath. Cut the center lath and notch out space on the upper and lower lath so that the box mounting brackets can rest on the uncut portions of the end laths. Before screw fastening the brackets to the lath make a notch under the receptacle mounting screw holes at each end of the box. This will provide clearance for the receptacle screws. Courtesy of General Switch Company, Middletown, N.Y.

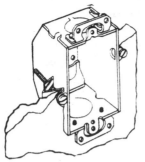 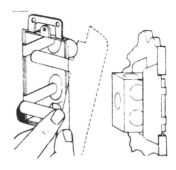

Fig. 5-12(b). "Grip-Tite" box clamps quickly in place. See how easy it is to anchor box securely front and back. Push box with connected cable in opening so that front brackets fit against wall. Then tighten side screws to bring side brackets up snug against wall. Courtesy of Sears, Roebuck and Co.

Fig. 5-12(c). Metal box supports can assure a stronger job. Insert supports on each side of box. Work supports up and down until they fit firmly against inside surface of wall. . . . . bend projecting ears so they fit around box. Courtesy of Sears, Roebuck and Co.

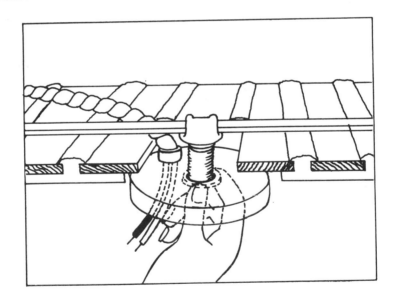

Fig. 5-12(d). If it is not practical to install a regular octagonal box for a ceiling fixture, use a shallow round box as shown. Slip the hanger through the hole, position it across the hole and fasten the shallow box to the stud to clamp the box against the hanger. Make any cable connections to knockouts in the top of the box. Wire nuts or other splices to fixture wires must then be made inside whatever fixture is fastened to the shallow box.

Two methods of holding a box to plasterboard or a ceiling fixture to a plasterboard ceiling are also shown in Fig. 5-12. One method makes use of support bars that slip into the plaster hole at each side of the box, while the other makes use of take-up clamps mounted to the sides of the boxes. The ceiling fixture is fastened on a shallow box with a special old-work hanger. Other methods are also used depending on the actual hardware you purchase.

From the previous discussion you can gather that the location of a box on the wall is pretty much a question of standard dimensions, the geometry of your house structure, and the location of receptacles as in your plan. However, how deep you set them in the wall is a different matter; the code is very specific here. If the finished wall or ceiling is a noncombustible material, such as plaster, the front edges of the box (this is the open front of a box on which you will mount a receptacle or switch) must be in a plane parallel with the outer wall surface and recessed not more than ¼ in. below the wall outer surface. In the case of combustible material, such as plywood panel, the front edges of the box must be flush with the outer surface of the wall. This point is important and can make or break you with the inspector. Note that with either installation, switch or receptacle plate covers fastened to the front of the box will tend to contain any danger of fire if an electrical malfunction and arcing occurs within the box.

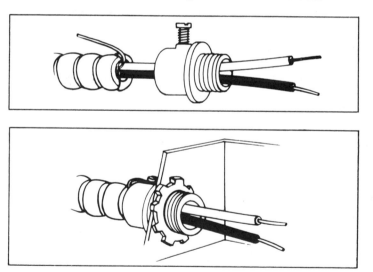

Fig. 5-13(a). Slip conductors through middle of BX cable connector, and slide connector over the armor, with grounding wire crimped in between. Tighten locking screw securely against armor. Fiber bushing should be against the front of the connector. Connect ground wire under the head of the locking screw before tightening the screw. Place connector into side of panel or junction box and secure tightly on inside with locking nut. Approximately six inches of conductor is left for making connections. Connect ground wire. Courtesy of General Switch Company, Middletown, N.Y.

Fig. 5-13(b). In lieu of capturing the ground wire under the setscrew of the connector, the non-metallic cable ground wire may be bonded to the box by a clip-on clamp as shown. Alternatively it may be connected to the box by a screw fastened into any wall of the box. Most boxes are equipped with tapped holes for the purpose. Courtesy of Sears, Roebuck and Co.

BOX CONNECTIONS

So now the boxes and fixtures are in place, cable or conduit is installed, wire terminations are prepared and ready for connection within the boxes to receptacles, switches, and fixtures. How is the cable or conduit fastened mechanically to the boxes so that you can close up and hopefully have no problems with the box? Remove the appropriate knockout patch using a drift or old screwdriver and hammer, then twist off the patch with pliers. Be sure that you know which openings you need and don't open any more than necessary. This is where the wires enter the boxes, but it is not the whole story. The conduit or the cable, both metallic or BX, must be fastened securely to the box at the knockout hole.

As with so much else in the electrical installation, there are so many types and arrangements for conduit and cable clamps that we could not possibly cover them all. The most common types are shown in Figure 5-13. In general, these connections have two things in common: they provide a firm mechanical link, a fixed clamp of some sort, that holds the BX, nonmetallic cable or conduit to the box, and a ground path is provided through the BX armor and ground wire, the conduit housing, or the nonmetallic cable ground wire back to the service entrance ground.

CLEAN UP AND PATCH UP

With everything wired in place and tested (see Chapter 12 for testing methods) your system is ready to function. However, there is one remaining step. All scars and cuts in the house structure must be covered with molding, plaster, cover plates, or the like.

Cover plates for receptacles and switches can be utilitarian or decorative. Boxes and fixture outlets should be covered with coverplates (available in wood, metal, porcelain, etc.) to match your decor and your pocketbook. Also, if you installed wiring behind baseboard molding, the molding must be replaced and faired into position. If you did not chisel a deep enough groove in the plaster, you might now have to chisel a coun017groove in the back of the baseboard or other molding. Try to cut deeper near the straps or other hangers so that the molding can fit flush against the wall through its entire length. Lastly, use special care in plastering around each box. The codes are specific about this and your inspector could get touchy. Use spackle or other patching mixture and fill all crevices between plaster and box so that when the coverplate is removed you do not see any gaping holes. Do not cover the edges of the boxes, because the inspector will want to ascertain that the boxes are not recessed more than the ¼ in. from the front wall surface.

6

SERVICE ENTRANCE

As we said in Chapter 3, we will assume you have or plan to install a three-wire, 115/230-V entrance supply. You know the service capacity in amperes. You will have three wires strung from the last utility company's pole to a specific drop-point on your house. These wires will fasten to the wall or roof peak and splice to three similar sized wires in a conduit or cable leading down the side of your house to an exterior or interior mounted meter. From the other side of the meter another three wires of the same size will run to the *service entrance distribution box*.

One brief word about the entrance wires before we continue. You should have one black wire, one red wire, and one white (neutral) wire in a service entrance cable. Large individual conductors installed in a service conduit are only available in black. It is necessary to use colored tape or paint on the ends to identify them. Wires must be sized to handle the service entrance current for which your distribution panel is designed (see the tables in Chapter 4 for ampacities). However, long entrance runs bring up the problem of voltage drop and require extra-size wires to handle the current. For example, a run that No. 6 wire would sufficiently serve for a distance of up to 75 ft would need No. 0 wire for a 400-ft run. Ask your power company to check this point for you, particularly if your house is some distance from the nearest road. Also, check these connections; they will probably be different from any others in your system. These heavy wires are joined with split-bolt solderless connectors (see Fig. 5-6.) These connectors are not self-insulating and they must be covered with insulating tape after connection.

GROUNDING THE SYSTEM

Before getting to the service entrance panel itself, let's discuss grounding again. As we have said, the system neutral wire must be grounded and the system ground (or bonding) wire must be grounded. These connections are made inside the service entrance panel and we cover that later. First though, let's explore the actual connection to ground outside your house and then how that connection is extended into the service entrance panel.

According to the NEC, a metal water-supply pipe must always be used to ground the system neutral bus. In addition, other grounding electrodes (see NEC, section 250-81) must always be used if available. When a water pipe and the supplementary electrode for a water-pipe ground cannot be used, an alternate method of grounding is to simply drive a copper pipe or copper-clad steel rod vertically into the soil outside the house foundation wall (at least a 3/4 in. diameter pipe or 5/8 in. diameter rod with an 8-ft minimum length). However, this is not always completely effective. The resistance from soil to ground rod should be less than 25 Ω (local codes sometimes call for less) and this is sometimes hard to achieve. When the resistance is more than 25 Ω, an additional rod must be driven at some distance from the first and connected with a buried, low-resistance bonding conductor to the first rod (see NEC, section 250-84). To establish a good ground, you must have a low-resistance area completely surrounding the ground rod. This includes both contact resistance with the soil and resistance of the soil itself. The rod must be free of paint or grease to minimize its contact resistance.

Resistance of the soil itself is the difficult part. Tests have shown that soil resistance within a 10-ft radius around the rod has an effect on the resistance between the rod and ground. For this reason, if you decide to lower ground

Fig. 6-1(a.) Megger test set for measuring ground resistance. (James G. Biddle, Co. photo.)

resistance by driving additional ground rods, all rods must be separated from each other by at least 20 ft. Then, when the rods are connected together by a buried bonding conductor, the soil resistance will be reduced because the resistances are electrically in parallel with each other. Three ground rods in an equilateral triangle with 20 ft to a side should make an effective system even if soil conditions are such that individual rod resistance runs high.

Soil conditions cause wide variation of ground resistance values. Lowest resistance occurs in soils made up of refuse, such as cinders, ashes, and brine water; these soils may show resistances below 15 Ω Clay, shale, and loams, on the other hand, average 24Ω, while mixing with sand, gravel, and so on can increase resistance to more than 90Ω; clearly too high. Soils of sand, gravel, and rock with no organic solid matter whatever will show still higher resistances up to more than 500Ω and you will probably need the help of an expert to assure the low resistance level required for an approvable ground.

Moisture and temperature also affect the resistance of soil. As you might expect, in wet weather the ground resistance will normally be half as much as during a dry summer. Similarly, cold and soil freezing increase resistance substantially. In fact, you must assure that your ground rod goes at least two feet deeper than the lowest frost line you can expect in your area. If necessary, resistance can be lowered substantially by treating the ground surface around the rod with rock salt or copper or magnesium sulphate and watering it throughly to soak the salt into the soil. This, however, requires renewal each year and periodic replacement of the rods if chemical action corrodes them.

There are several methods of measuring the actual ground resistance. Some are so cumbersome and unreliable, though, that we only recommend measurement with a megger® test set, which is a special type of ohmmeter [Fig. 6-1(a)]. If you cannot borrow or rent one, your local power company may be willing to assist you.

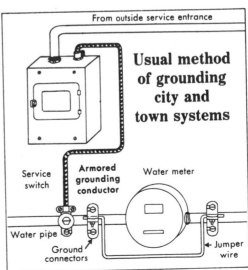

Usual method of grounding city and town systems

Fig. 6-1(b). Grounding conductor clamped to water pipe. (Courtesy of Sears, Roebuck and Co.)

Figure 6-1(b) shows what is probably the best possible ground you can have. If your water supply enters the house through metal piping and is nearby to your service entrance, clamp a grounding conductor onto the water pipe on the input side of the water meter. This conductor should be no smaller than No. 8, or 1/2″ wide, if flat braid is used. It may be bare or enclosed in armored cable if you desire. Clamps of standard commercial type that will ensure a mechanically secure, low-resistance contact must be used.

SERVICE ENTRANCE PANEL

When you have made a firm, low-resistance ground connection to a water pipe or ground rod, your next step is to connect this ground to your service entrance panel.

You can get either flush-mounted or surface-mounted panels and they are normally mounted about shoulder height and as near as possible to the supply wire entrance point. As shown in Fig. 6-2, most service entrance panels have two front doors. The first door covers about half of the panel front surface. When it is swung open, the circuit breakers, which are insulated and safe to touch, are visible. However, no wiring should be visible or accessible. You can safely turn off all system power or power to any individual branch circuit when this door is open. However, you should still be careful, particularly if your floor is wet. As a precaution, stand on a few dry 2 x 4 beams or a plywood panel whenever you open your service entrance box.

The second door is generally the entire front surface of the box. You will

Fig. 6-2. Service entrance panel. (Graf–Whalen photo.)

probably find it screw-fastened at all four corners. When the screws are removed the front cover of the box (containing the smaller door) can be swung open. Now we have a different situation; Many wires and terminals, hot wires as well as grounds, are visible and touchable. Before you ever open this cover you should always turn off the power to the system or you are asking for trouble. In a circuit breaker box, set the main breaker switch to OFF; in an old-fashioned fuse box, remove the main supply fuses.

The 230-V, three-wire supply will enter the box from above and connect with solderless terminals to the two hot and one neutral terminals. These supply wires will always be hot unless the utility company turns off power at the meter or elsewhere outside your house. The main breaker only shuts off power to the branch breakers, not the supply buses. Therefore, remember not to touch these wires at any time when the panel is opened.

Whether your panel is fitted with fuses or circuit breakers, it should have a ground bus to which your ground wire must be connected. A neutral bus is a solid conductor (usually a metal bar) with many screw terminal points for attachment of conductors. Between these, there is close to $0\ \Omega$ resistance. Therefore, if a ground wire with $25\ \Omega$ or less to ground is connected to the bus, any other wires connected to the bus will also be almost exactly at the same resistance above ground. It is this bus point that is ground or common as far as your house wiring is concerned.

We have already said that all neutral conductors of all branch circuits must be connected to ground and that ground (or bonding) wires of each branch circuit must also be grounded. You will find that service panels generally have two buses, one for connection of the external ground connection and all neutral wires (including the heavy supply neutral) and one for connection of all ground (bonding) wires. If you look closer, you will find that both these buses are, in turn, connected together so that in effect you have one solid ground bus for both the neutral wires and ground wires; all wires connected to the terminals of either bus are therefore at an identical resistance with respect to ground.

In your service panel, both ungrounded wires (normally red and black) will enter from the top and connect to two separate hot buses through the main circuit breaker, fuses, or switch disconnect arrangement. If you connect an a-c voltmeter from red or black to neutral (white or bare), you should get a reading of 115 V. The same meter from red to black, though, will show 230 V.

In a circuit breaker box, the main disconnect will be a circuit breaker; in a fuse-type box, it can be an auxiliary lever-operated switch in the supply line or, more likely, a pull out fuse block such as shown in Fig. 6-3. When the block is in place, contact is made through the fuses to provide current to the two hot buses. When the block is removed, all current to the branch circuits is cut off. Again, remember that fuse block removal (or main breaker turn-off) only cuts current in the branch circuit buses; the supply lines are always live unless power is shut off by your utility company.

Fig. 6-3. Pull out fuse block. (Graf–Whalen photo.)

Fuse-Type Service Panel

Let's discuss Fig. 6-4, which is a half-pictorial, half-schematic view of a fuse-type service panel. You can identify at lower left a ground connection to the ground block, grounding all terminals and wires connected to the bus. At the top, the heavy red, black, and neutral supply lines enter the block, including the neutral bus and all neutral wires. (Cartridge fuses are designated 1 thru 4 in our illustration for ease of description.) Red is connected to main fuse 1, black to 2, and white to the neutral bus. Fuses 1 and 2 carry all the current that enters your system; therefore, they must be of high-current capacity. For a 100-A/230-V system, each fuse will be 100 amperes. In many panels they will be mounted in a removable fuse block, as shown in Fig. 6-3. At the other side of the main fuse, each hot line splits into several branches. In our example, the current through fuse 1 goes to cartridge fuse 4, and to two plug fuses on the right and an unfused terminal. Fuse 2 supplies fuse 3, two plug fuses on the left, and an unfused terminal.

Taking the cartridge fuses first, fuses 3 and 4 will normally be mounted in a separate fuse block, similar to 1 and 2, and will normally be used to supply 230 volts to an electric range. The range supply cable (three-wire with ground) exits from the box toward the right in our example. You can see that whereas removing the left-hand fuse block (fuses 1 and 2) cuts off current to the entire house system, removal of the right hand block (fuses 3 and 4) will kill only the range branch circuit. Also, note that although fuses 1 and 2 are sized for the capacity of the entire system, fuses 3 and 4 are generally of a lower rating, possibly 30 amperes each, to handle the range current only.

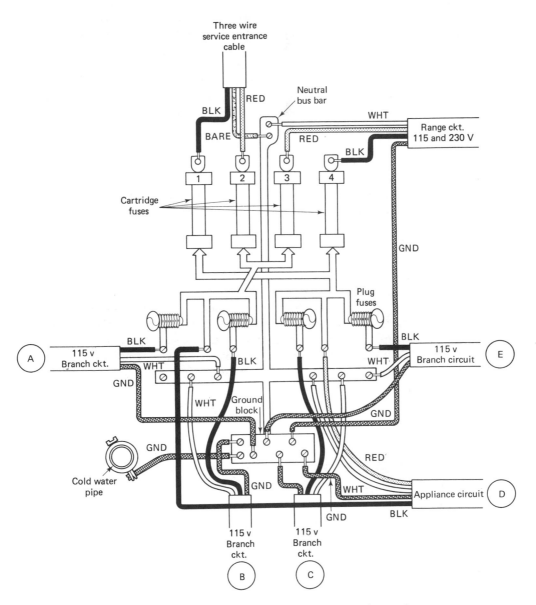

Fig. 6-4. Pictorial and schematic of a fuse-type service panel.

Five circuits are shown connected to the box. We have labelled them A thru E. A and B are two-wire with ground circuits. Each taps 115 V from one of the plug fuses on the left. Neutral and ground wires of each are connected to the respective buses. Circuits C and E are similar, taking 115 V respectively from the two right-hand plug fuses.

Circuit D, however, is different from the other four. Can you see the difference? First, this is a three-wire with ground cable. The black wire connects to one unfused terminal on the left and the red wire to an unfused terminal on the right. They each receive 115 V and therefore, as with the range circuit, there are 230 V between the black to red wire. This cable can be used to supply an appliance circuit that may need 230 V, for example, a clothes dryer, in which case it will have to go through a fuse-protected switch (separately housed) next to the main panel before reaching the appliance. Also, it can lead to an auxiliary fuse box with two, four, or more fused circuits branching from it. In either case, the cable has to be sized to handle the probable current draw. This type of circuit sprawl is most common in old work, where an original system has gone through a series of updates each time an appliance was bought.

On the back of the hinged cover of most service panels is a diagram showing how the panel is wired and how you should wire each house circuit to the terminal screws provided in the panel. You will probably be able to recognize similarities between our diagram and yours. Even though details may differ, the principles are the same in all. Adding a circuit in a fuse panel, such as the one Fig. 6-6, is relatively simple if there is an unused plug fuse or empty unfused terminal to which you can attach an auxiliary fuse box. If everything appears full, however, your best bet will be to break into an existing circuit with an auxiliary fuse box, such as the one shown in Fig. 6-5. Remember, though, if you add several circuits where there was one before, supply wires into the auxiliary box will have to be

Fig. 6-5. Auxiliary fuse block. (Graf–Whalen photo.)

sized to handle the multicircuit load. In time, you will probably want to replace the sprawling fuse-type service panel with the more flexible and neater circuit breaker panel.

Circuit-Breaker-Type Service Panel

You can add auxiliary fused circuits for awhile but inevitably, when the connections become too complicated and the space in the box gets too crowded, your best bet will be to add an entirely new panel, and this will have to be a circuit breaker panel in accordance with the latest NEC requirements. You will find that these panels are neater and cleaner looking. They are also more compact and can accommodate a large number of circuits. A typical panel will handle 200 A of main supply current and supply ten 230-V branches, twenty 115-V branches, or any combination in between (e.g., three 230-V range, clothes dryer, air conditioner circuits, etc. and fourteen 115-V general-purpose and appliance circuits). Bus connections are simple, as in the fuse-type panel. Branch and main breakers install easily and there is more space for expansion.

To install a branch circuit you can work from either end of the line. In new work, before the utility company turns on power to your system, you can connect a cable to the panel then work progressively along the circuit to switches, outlets, and fixtures (see Chapters 7 and 8). In old work, however, you would be better off installing the entire branch circuit, outlets, and so on, then make your connection to the panel last so that power to the entire system will be off for a minimum length of time.

BRANCH CIRCUIT CONNECTION TO SERVICE ENTRANCE

For simplicity of description though, we will assume that you are doing new work and the utility company has not turned on the power yet. A typical circuit would be installed as follows.

With the service entrance cover off (see Fig. 6-6), select the circuit breaker terminal to which you will connect a 115-V branch circuit cable. The service entrance wiring diagram, usually mounted inside the front cover, and the route of the cable outside the box will help to determine the general location you want. Now note the closest knockout in the sidewall of the box for cable entry. Measure from this knockout location to the ground and neutral buses, often near the bottom of the box. This will show what length of cable outer sheath must be removed for the wiring connections inside the service entrance.

You should find many knockouts in the sidewalls of the box for maximum flexibility in wiring arrangements. *However, do not open any knockout unless you are sure you will use it.* (Some service entrance boxes are designed for mounting between wall studs so that they may be recessed in the wall. This type will not have knockouts in the sidewalls, but in the top and bottom.)

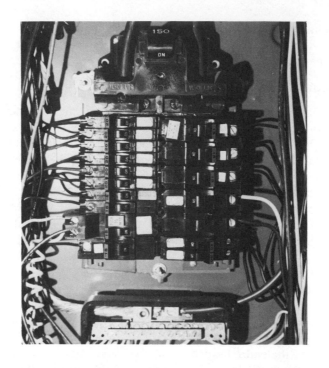

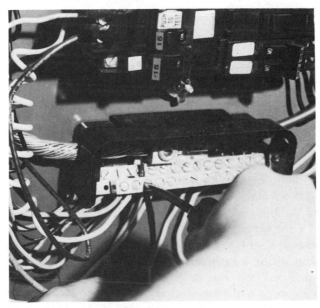

Fig. 6-6. Connection of branch circuit to service entrance. White wire connects to neutral bus, and black wire to circuit breaker. (Graf–Whalen photos.)

Assume that NM14-2 with ground cable, one of the most common types (see Chapter 4), is used. If it is 10 in. from the knockout to the buses, remove about 14 in. of outer sheath from the cable end to expose this length of the three wires (one black, one white, and one bare or green ground wire). Remove the knockout to allow entry of the cable. Find the small sections of the circular knockout perimeter where the hole is not cut through the wall. Place a sharp metal tool against the knockout about midway between the uncut section and give the tool a sharp rap with a hammer. This will partially twist the knockout disc so that you can grab it with pliers and pry it out, leaving a neat, circular hole that will accommodate a cable clamp.

From the outside, insert the threaded shank of a cable clamp through the knockout hole. Fasten the clamp locknut to the shank inside the box and finger tighten the locknut against the inside wall. The locknut has notches around its perimeter. Place a screwdriver or similar blade into these notches and pry or hammer tap the locknut for final tightening against the wall.

Now insert the three wires of the cable through the center of the clamp from the outside. Feed the cable in until the cable outer sheath rests in the clamp and the exposed wiring is all inside the box. Fasten the clamp to the outer surface of the cable sheath; most clamps have a lock screw arrangement on the outer surface for locking the cable.

Dress the ground wire from the knockout to the ground bus in a smooth path. You should find a series of terminals on the bus, usually screw or bolt type clamps, for the wires. Cut off any excess length of the ground wire that might cause clutter in the box, and fasten to a convenient terminal. Fasten the clamp firmly. Note that if there are not enough wire terminals for the grounds of all the circuits you are adding, a separate ground bus can be purchased and mounted in pretapped holes. Also, if the ground wire is a green insulated wire, remove about 1/2 in. of insulation and clamp the exposed end to the terminal.

Now, connect the white, neutral wire to a terminal on the neutral bus in the same way as just described. Connect the black, hot wire to the circuit breaker terminal you have selected. You may have to shorten the black lead a bit for smooth connection because the connection point may be close to the knockout.

To install a 230-V branch circuit, use a three-wire-plus-ground cable (black, red, and white, and ground wire), connect the ground and neutral wires as just described and connect the black and the red wires to two hot terminals on different buses, providing 115 volts in each hot lead.

The circuit breaker is now inserted in the unused position adjacent to your connection or connections. As we describe in Chapter 4, breakers are available in different voltage and current ratings as well as physical types. First, you want a breaker that will fit the panel correctly; check carefully with your supplier. Some panels simply have open cavities into which the breakers can be snapped, while others have positions covered by oblong knockout patches that must be removed before breaker insertion.

Next, select the type of breaker required to handle the voltage to be supplied to your new circuit. A single-pole breaker will give protection to a single 115-V circuit and occupy one space in the panel. A double-pole breaker will protect a 230-V circuit, occupy two spaces, and is supplied by both black and red wires. Finally, a single-pole double breaker occupies one space, protects two lightly loaded 115-V circuits, and also takes two hot line connections and two neutral connections. Two two-wire with ground cables are required for this type of installation.

Your last selection requirement is to choose a breaker that is rated for the current you intend to draw from the circuit. As we said, general-purpose circuits will usually be rated at 15 A and require 15 A circuit breakers; appliance circuits will need 20 A breakers for protection. Naturally, the cables in these branches must be sized to handle the load. You now have the cable connected at the service entrance panel, properly protected by circuit breakers, with cable wiring sized to handle the expected load. Now, move out from the panel to install the circuit in accordance with your plan, outlet-by-outlet (Chapter 7), switch-by-switch (Chapter 8), lighting fixture-by-lighting fixture (Chapter 9).

7

BRANCH CIRCUITS

Adding Outlets – Adding Circuits – Grounding

If you learn only how to wire branch circuits containing direct or switch-controlled ceiling outlets and dual receptacle outlets for extension cords, you will probably have learned how to do the bulk of all the wiring you will install. This sounds simple, and it is, but there are many variations to be covered and it will take a while to get through them. Switch-controlled circuits are covered in Chapter 8, but outlets are discussed in the following sections. Roughly speaking, after we discuss the general rules of wiring branch circuits, we will cover the subject of outlet wiring in the following sequence:

1. Adding a single new outlet and breaking into an existing branch circuit.

2. Wiring a branch circuit to any number of outlets in a string (up to the maximum safe number for the wiring). This would be typical of the wiring for a new addition such as a bedroom, bathroom, den, or attached garage.

3. Wiring a number of outlets through junction boxes, so that, although they branch out like a family tree, electrically they all will be wired in parallel.

4. Wiring a branch circuit dedicated to one outlet. This type of circuit is normally used for large appliances such as an electric clothes dryer or range. Note that some codes allow connection directly to the appliance rather than to a dedicated outlet and this variation is covered.

5. Wiring a GFCI into a branch circuit to protect the circuit.

Before we start, let's get some of the general rules of wiring clearly established. From there on we will simply state what to do and you can understand why each step is taken.

As you know by now, each branch circuit has one or two hot leads not connected to ground and a neutral lead that is grounded at the service entrance. We call the ungrounded lead(s) hot to indicate that they are dangerous to touch; remember, though, that electricity flows through both the hot and neutral leads. Most of the working components of the circuits (outlet receptacles for lamp and appliance plugs, light fixtures, or wired-in appliances) are connected between the hot and the neutral wires and are therefore wired in parallel with each other. You could represent this arrangement by the diagram shown in Figure 7-1.

As we also said before, the circuit overcurrent protector and switch controls are only installed in the hot wire. Therefore, Figure 7-1 could be redrawn as shown in Figure 7-2.

Now that we have that established, how do you keep track of the hot and neutral wires to be sure that you do not cross yourself up? This is where color-coding is important. Given a cable such as type NM 14-2 with ground wire (see Fig. 4-3), you will find one wire covered with black insulation, a second with white insulation, and a bare ground wire. Use the black wire only for hot leads and the white only for neutral. Then, throughout a branch circuit, whether it is a straight cable run from the service entrance panel to a dedicated, wired-in appliance or a branch circuit that separates into many offshoots, connect black wires only to black wires and white wires to white wires and your wiring will be correct almost automatically. There are exceptions to this, as in wiring switches, but we cover them at appropriate places later.

The bare grounding (or bonding) wire must be treated slightly differently—all cable ground wires entering a box must be connected to each other, to the box, and to the ground lug of a grounding-type receptacle, if used. Various combinations are possible. You might think that a box could be grounded by connecting the bare ground wire to the green-tinted ground screw terminal of the

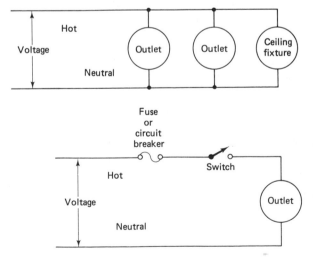

Fig. 7-1. Parallel wiring.

Fig. 7-2. Diagram showing circuit breaker and switch control in the hot wire.

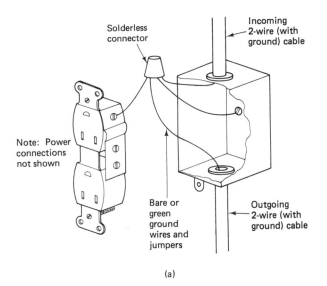

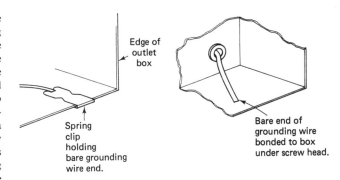

Fig. 7-3. (a) Even though the receptacle grounds to the box through the mounting screw, a separate ground jumper wire from the green receptacle terminal to the box is also required. Connecting in one solderless connector, the incoming and outgoing cable grounds with the two ground jumpers (one to the green terminal and the other to the box) assures a good ground bond. The ground wires may be bare wire or green wire with the ends bared. (b) Typical methods of connecting ground wire or ground jumpers to the outlet box.

receptacle. This screw provides electrical continuity to the receptacle mounting ears so that when the receptacle is screw-fastened to the box, the box has a path to ground and is protected. However, this type connection could give poor contact. Besides that, codes require that the ground contact must not be broken when the receptacle is removed. Therefore, we recommend any of the methods shown in Figure 7-3. All grounding wires may be twisted together and fastened to the box by means of a sheet metal screw or connected together by wire nut and connected by a single jumper* to the sheet metal screw or wire clip fastened to the box (see

*A jumper wire is normally a short piece of wire, often bare wire, connecting two points together electrically.

Fig. 7-3). This method allows you to fasten two, three, or four wires to the box without trying to fit them all under one screwhead. Also, because, with a grounded circuit, you will be installing a grounding-type receptacle, a separate jumper wire is connected from the green-tinted ground screw of the receptacle to ground. This jumper can be a bare conductor or a green-insulated wire with the ends stripped. Do not use any other colored wire. Connect the jumper to the box with a sheet metal screw or by fastening it by a wire nut to the other ground wires or by a ground clip fastened to a front edge of the box. Using the clip, trap the bare end of the jumper under the clip so that the wire presses against the box wall.

All circuit components are connected from the hot to the neutral line and are in parallel with each other. This can be seen quite easily for outlets placed one after the other in a chain; actually, this is a convenient way to wire them. But it is also true that components are in parallel on a circuit that branches at junction boxes like a family tree. It will look more complicated but it will still be identical to Figure 7-2 electrically.

Another rule that we have mentioned but must repeat for emphasis is as follows: All 115/230-V wiring connections, wire to wire in a splice or wire nut, or wire to terminal of a receptacle or switch, *must* be made inside a junction or outlet box. There are *no exceptions* to this rule.

To continue the wiring color-coding rules; connect black wires only to copper (or brass) terminals of a receptacle or switch and white wires only to the silver (or chrome) terminals. This is essential to polarize the receptacle blade arrangement so that we can control which line cord wire is hot and which is neutral. Consider an older line cord of the type with which we are all familiar. The plug blades are identical. The plug can be inserted into the receptacle either way. In this case, you do not know which wire is hot and which is neutral; you do not know which wire should go to the lamp or appliance switch and which to the neutral side of the load. This is undesirable and not as safe as the polarized aproach, now universally required.

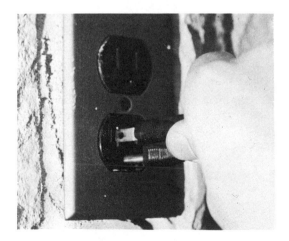

Fig. 7-4. Do not install grounding type, 3-blade receptacle in old work. (Graf–Whalen photo.)

In the discussions that follow, we assume that you will be installing grounding-type receptacles in branch circuits with grounded devices, in other words, circuits with all outlets, junctions, and switch boxes connected by bonding conductor back to the entrance ground. This assumption is essential because it is the only new work allowed by the NEC. As we discussed in Chapter 3 on planning your system, you have to determine whether your existing circuits are bonded to ground. If they are not, the only allowable work is replacement of the existing two-blade receptacles or switches. You must *not* install a grounding type, three-blade receptacle (see Fig. 7-4) in old work because it would be dangerously misleading. For the same reason, you may not extend these circuits by adding outlets or switches because all new receptacles must be of the grounding type and grounding-type receptacles may not be added to ungrounded circuits. If you want to add outlets, your only recourse is to add entirely new grounded circuits. Besides being required, it is the safer, modern method and does allow you to plug in appliances with grounded housings and three-blade line cord plugs. Naturally, grounding receptacles will also accommodate the standard two-wire lamp cord plugs, so they are versatile as well as safe.

ADDING A NEW OUTLET

Let's start off easy. Say you want to add a single new outlet with dual receptacle just for lamps. The mechanical operations of installing the outlet box, stripping, and connecting the conductors are described in Chapter 5, while the hardware to use and criteria for cable selection and branch circuit capacity calculations are covered in Chapters 3 and 4.

Now to get down to specifics. For a lamp outlet, you will probably use a Type NM 14-2 with ground wire cable (Fig. 4-3.) You know where you want the outlet and you have to find the nearest, most convenient connection point for your cable. The most likely prospects are the following:

1. The nearest outlet in the room, if it is at the end of a cable run and has only one cable entering the box.

2. The same as 1., but connecting to a dual receptacle in the middle of a circuit.

3. A junction box in a basement or attic to which you can snake your cable easily.

4. A length of cable into which you can cut. You can then connect your new cable as a side branch by means of two junction boxes.

The first step is to turn off the power to the area on which you will be working. You *must* turn off power while working at the service panel, so do that last. While working on an existing branch, either throw the circuit breaker for that branch to OFF or unscrew the fuse and put it into your pocket for safe keeping. To avoid having a branch out of service for an excessive length of time, do the new outlet end first.

Cut the outlet opening at your selected location and run the cable from the opening to the most convenient point of the selected branch circuit using fish tape (Chapter 5) or other methods, as required. At the new outlet location, remove the plastic sheath from the last 6 to 8 in. of the cable and strip the wire ends for connection to the receptacle terminals. For ease of installation, use the deepest outlet box that will fit your location. Knock out the appropriate patch of the outlet box, insert the wire ends (including the bare ground wire) through the hole and out the front of the box, clamp the cable to the box at the knockout hole, and install and fasten the box in the opening you have prepared. Connect the wire ends to the dual receptacle using either screw terminals or push-in terminal holes depending on hardware you are using.

> **NOTE:** *For screw terminals, strip the last 1 in. of each wire. For devices with push-in terminals, strip as required by your hardware.*

Connect the cable ground wire to the outlet box and connect a jumper from the receptacle ground screw to the box. Carefully press the receptacle into the outlet box and fasten in place with captive screws on the receptacle. Press the wires into the box behind the receptacle. It helps to twist the wire into half-loops using a needle-nose plier. The wires can then be tucked into the box more easily.

Assuming the cable has been strung to the selected branch circuit connection point, it will probably be one of the four types of connection point mentioned previously.

Connecting to a dual receptacle at the end of a branch (see Fig. 7-5) is the easiest method so let's discuss that now.

First, be sure to turn off power to the outlet.

Remove the cover plate of the box and loosen and pull the existing receptacle from the box without disconnecting wires. Break away plaster or remove panelling around the box to allow fishing and fastening the new cable to the box. Strip the cable-end wire insulation as required. Knock out a convenient patch for the new cable to enter the box, insert the wire ends through the hole and clamp the cable to the box. Now how does your new receptacle get its electricity?

Remember that dual receptacles have two screw terminals for hot leads (copper or brass) and two for neutral (silver or chrome). Connecting a hot wire to either brass screw terminal connects both the hot terminals to the circuit and energizes both receptacles; the same is true for neutral screw terminals. If the receptacle is already connected to a branch circuit via the screw terminals, you might simply connect your new wire to the unused screw terminal (black to copper and white to silver) and your new outlet will become part of the branch circuit. You might except for the NEC. The NEC will not allow this type of connection for the neutral wire because one bad screw connection could destroy the neutral for a whole chain of outlets, clearly a dangerous condition. Therefore, make the neutral connection for each outlet by means of a wire nut. Connect the two neutral wires in a wire nut along with one end of a short, white jumper wire.

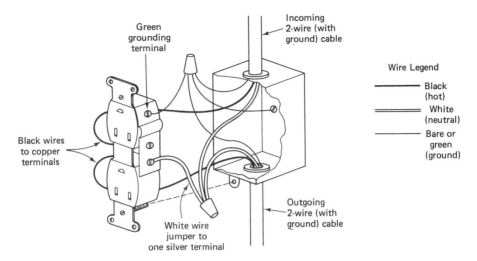

Green grounding terminal

Incoming 2-wire (with ground) cable

Wire Legend

——————— Black (hot)

═══════ White (neutral)

——————— Bare or green (ground)

Black wires to copper terminals

White wire jumper to one silver terminal

Outgoing 2-wire (with ground) cable

Fig. 7-5. Connecting to a dual receptacle at the end of a circuit branch.

Connect the other end of the jumper to either the silver (or chrome) terminal. Note that you could connect the black wires in the same way, with a jumper to one of the copper terminals, but the NEC does not require it. A break in this circuit due to a bad screw connection would be a nuisance but it would not be dangerous. Instead, for convenience, you can connect the two black leads to the two copper terminals as shown in Figure 7-5. In a grounded circuit the existing cable ground wire should be connected to the box and to the receptacle. Disconnect these wires and reconnect them with the end of the ground wire from the new outlet. Fasten all ground wires together by a wire nut and connect a single jumper to the box by a fastening screw as shown in Figure 7-5. Then carefully tuck all wire loops back into the box as you reinstall and fasten the receptacle to the box.

Turn on the branch circuit power. It is a good idea at this point to check the functioning of all connections by plugging a lamp or test light into both receptacles of both outlets. Check for proper functioning, then install cover plates on each outlet and the job is done.

The second junction type mentioned previously is somewhat more difficult. With the new outlet installed and cable run to a dual receptacle where all screw terminals are occupied, connection depends somewhat on the size of the existing box or a box you may need to substitute for the old box if you have the room. Assuming that you have the room, disconnect one set of cable leads from the receptacle's copper and silver screw terminals. Extend your cable termination into the box, clamp the cable to the box, and connect the cable as follows.

Prepare three short jumper wires, one with black insulation and stripped ends, another white with stripped ends, and a third, bare or green with stripped ends. Twist one end of the black jumper with the black wire of your new cable and with the black wire you disconnected from the receptacle. Fasten the three with a wire nut and fasten the other end of the black jumper to one copper screw

119

terminal of the receptacle. Do the same for the white wires, fastening the white jumper to one silver terminal of the receptacle, and the same for the bare bonding conductors, fastening the bonding jumper to the outlet box wall with a sheet metal screw. You should now be able to restore the box to the wall and turn power on again. Check the installation for proper functioning. If all is well, install the cover plates on each receptacle.

Before we get into junction types three and four, we would like to digress and discuss installation of entire circuits, including junction box branches. The technique will probably soon become obvious. Because this discussion has been detailed, it has already covered many points that you will need to know about your circuitry. From here on it gets even easier.

Wiring a Straight-Line Branch Circuit

Electrically, a branch circuit with a chain of dual receptacle outlets is the simplest circuit to wire. Whether there be two, three, or even more dual outlets on the circuit, they are all alike and are wired the same as the outlet shown in Figure 7-5. The incoming and outgoing cables are clamped to the box with wire ends extending out through the open box front. The incoming black wire is connected to one copper terminal screw and the outgoing black wire to the other copper terminal screw on the receptacle as shown.

Both copper terminals are identical points electrically. The incoming and outgoing white wires are twisted together with one end of a white jumper and fastened with a wire nut. The other end of the jumper is then fastened to either silver terminal to provide the neutral connection. As described earlier, the ground wire should be twisted together, along with a jumper wire from the ground terminal of the receptacle and a jumper wire to the box. These should all be connected together securely with a wire nut. The jumper to the box is then screw-fastened to the box wall. Repeat this wiring procedure for each outlet in the chain to provide a positive electrical ground connection for each outlet box.

Wiring a Branching Branch Circuit

If, for example, you want to split a circuit into two branches to extend around two sides of a room, you must use a junction box as shown in Figure 7-6. In the junction box, the incoming supply cable is connected to two divergent cables by means of wire nuts. The three black leads are twisted together and fastened by a wire nut, then three white leads are fastened together. The three bare bonding conductors are twisted together with a bare or green-insulated jumper, fastened together by a wire nut and electrically connected to the junction box wall by means of a sheet metal screw that fastens the other end of the jumper.

Each outgoing branch cable may supply a chain of outlets as described previously or be branched further by junction boxes into additional chains of outlets. No matter how many branches are constructed, though, each outlet will

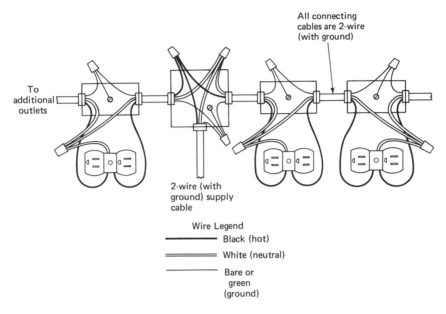

All connecting
cables are 2-wire
(with ground)

To
additional
outlets

2-wire (with
ground) supply
cable

Wire Legend
———————— Black (hot)
============ White (neutral)
———————— Bare or
             green
             (ground)

Fig. 7-6. Splitting a branch circuit into sub-branches is accomplished using a junction box as shown. Each side branch may also be branched again in the same way.

be in parallel with each other one in the circuit and will be across the line from hot to neutral wires, as long as you follow the wiring rules (black to black and white to white).

Connecting an Added Cable into a Junction Box

As we promised, we can now return to our first example—adding a new outlet and breaking into a circuit by means of a junction box. The third break-in method mentioned earlier requires finding a junction box with unused capacity. If only three cables branch out from the box, knock out a fourth patch, add your cable, and clamp it to the box, then connect the cable into the circuit using wire nuts as shown in Figure 7-7. You will have to replace the existing wire nuts with wire nuts that can hold four wires. Replace the box cover only after you have tested all receptacles for proper functioning.

A word should be said about box capacity. To accommodate four two-wire cables (in our example four NM 14-2 with ground), a junction box would have to be octagonal in shape, about 4 in. across and at least 1½ in. deep. This size box can handle 11 wires (not including ground wires). If it contains cable clamps or switch housings, subtract one wire from the permitted number for clamps or each device contained in the box (see Table 7-1 for additional details). Generally it is better to use wall cable connectors rather than buy boxes with built-in clamps because the wall connectors allow more space for your splices.

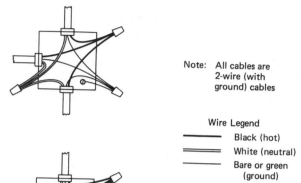

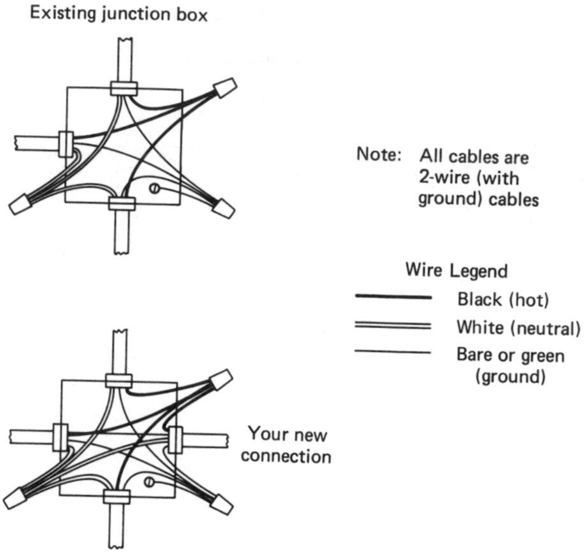

Existing junction box

Note: All cables are
2-wire (with
ground) cables

Wire Legend
——————— Black (hot)
═══════════ White (neutral)
——————— Bare or green
(ground)

Your new
connection

Fig. 7-7. Connecting a fourth cable into a junction box already containing three connected cables.

TABLE 7-1.   **Deep Boxes**

Box dimensions, inches trade size	Cubic inch cap	Maximum number of conductors			
		No. 14	No. 12	No. 10	No. 8
3¼ × 1½ octagonal.........	10.9	5	4	4	3
3¼ × 1½.................	11.9	5	5	4	3
4 × 1½.................	17.1	8	7	6	5
4 × 2⅛.................	23.6	11	10	9	7
4 × 1½ square	22.6	11	10	9	7
4 × 2⅛.................	31.9	15	14	12	10
4¹¹⁄₁₆ × 1½ square	32.2	16	14	12	10
4¹¹⁄₁₆ × 2⅛	46.4	23	20	18	15
3 × 2 × 1½ device	7.9	3	3	3	2
3 × 2 × 2................	10.7	5	4	4	3
3 × 2 × 2¼	11.3	5	5	4	3
3 × 2 × 2½	13	6	5	5	4
3 × 2 × 2¾	14.6	7	6	5	4
3 × 2 × 3½	18.3	9	8	7	6
4 × 2⅛ × 1½	11.1	5	4	4	3
4 × 2⅛ × 1⅞	13.9	6	6	5	4
4 × 2⅛ × 2⅛	15.6	7	6	6	5

TABLE 7-1 (Continued)   Shallow Boxes*

Box dimensions; inches trade size	Maximum number of conductors		
	No. 14	No. 12	No. 10
3¼	4	4	3
4	6	6	4
1¼ × 4 square	9	7	6
4¹¹⁄₁₆	8	6	6

*Any box less than 1½ inch deep is considered to be a shallow box.

Connection to an Intermediate Point of a Cable

The fourth break-in method for a new outlet also uses the junction box technique. As shown in Figure 7-8, the connection is accomplished by using two junction boxes. First, turn off power to the branch circuit involved and check carefully that it is the right circuit. You might even want to throw the main system switch off or remove the main fuse block to be sure. Cut the cable at the point you have chosen for connection; it should be a location where two junction boxes could be mounted easily. Strip both the cut wire ends and prepare them for connection. Note that you will not have enough slack for connection. Therefore, it will be necessary to install each cable in a separate junction box. Mount the boxes about a foot apart. Fasten a new short section of cable to one box and connect it to the wires within by means of wire nuts. Connect the other end of the short cable run to the second junction box, along with the cable end from your new outlet. Connect all black, all white, and all ground wires of the three cables together with wire nuts and connect a jumper from the ground wire splice to the box. Test all receptacles and all ground circuits for proper function before closing up boxes. Be sure to use covers and screw-fasten them to each junction box. Now let's continue with circuits that offer additional features.

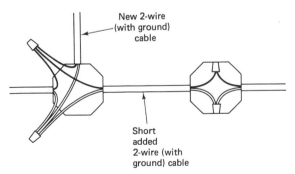

New 2-wire (with ground) cable

Short added 2-wire (with ground) cable

Fig. 7-8. Cut the cable at a convenient point. Then add two junction boxes about one-foot apart and short cable between them. Connect new cable to one of the junction boxes. Connect all wires and ground wire as applicable.

USING THREE-WIRE CABLE FOR TWO TWO-WIRE BRANCHES

Though we discussed it earlier, we will repeat the description of a three-wire cable to refresh your memory. A three-wire cable such as type NM 12-3 with ground contains a black wire, a red wire, and a white wire plus a bare ground wire. At the service entrance, the black wire is connected to one 115 V leg and the red wire to the opposite 115-V leg. The single neutral wire serves both 115-V legs because they are of opposite phase. Both the red and the black wires are protected by separate overcurrent protection devices. At the working end of the cable, therefore, 115 V is available from either the black to neutral wire or the red to neutral, or both simultaneously, as well as 230 V from the red to black wires.

The voltage characteristics of a three-wire cable can be used advantageously when you want two 115-V circuits at some distance from the service entrance, such as in an attic, attached garage, or workshop. One of the easiest methods is to run a single three-wire cable (saving the work and money of running two cables) to the remote location, then split it into two 115-V circuits, as shown in Figure 7-9. If you use one 14-3 wire, connect to two 14-2 wires for the separate circuits. From the picture and caption you can easily see how the split is made. However, you will note that this is one place where you encounter an exception to the black-to-black rule. At one splice you will be connecting black to red, which is perfectly sound electrically but potentially misleading and local codes have interpreted the

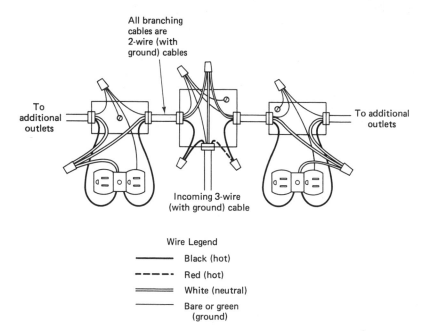

Fig. 7-9. Splitting to two-wire circuits from one three-wire cable. Note the solderless connection at the lower right of the central junction box. We connect red to black to provide 115 V to the right hand branch.

usage in different ways. Some local codes require you to paint the visible ends of the black wire red so that you know it is spliced to a red wire, while others ban it altogether, effectively eliminating the use of this wiring shortcut. You must check this point carefully to avoid the risk of it not being approved by your inspector.

We have talked of supplying two-circuits using NM 14-2 with ground cable from 14-3 with ground cable. You might think intuitively that the three-wire cable would have to be heavier wire to carry the load of two circuits. However, this is not true because the red wire carries the current for one two-wire circuit and the black wire carries current for the other. The neutral does not have to carry twice the return load either and does not have to be twice as large. What actually happens is that the current in the black and red wires are out of phase; rather than adding together in the neutral wire, they occur 180° out of step with one another. And so, the neutral wire carries current flowing through the load from the black wire at one instant and the current flowing through another load from the red wire during the next split-second. The effect is that the neutral carries only *one* circuit's current at a time.

SPLIT-RECEPTACLE WIRING

A second wiring advantage of three-wire cable is that you can wire your outlets as split receptacles, with one receptacle of a dual unit on one circuit and the other receptacle on another. For this purpose, you have to buy special receptacles that are made with a breakout section between the paired terminal screws on each side. The breakout is constructed so that it can be snapped off with a screwdriver or pliers to electrically separate the two terminals.

Split receptacles can be used in two ways. In place of two 15-A circuits, each with a chain of 10 outlets, you could install 20 split-receptacle outlets on one chain. Depending on your layout, the long chain could save considerable time and effort. Also, with a random distribution of appliances and lamps, you are not likely to overload either circuit. It would be a good idea to keep track of which outlet is on which circuit for troubleshooting purposes later on. In one room you might put all black wire receptacles on top and red on the bottom, then alternate it in the next room.

Figure 7-10 shows the electrical connections required for this circuit-splitting technique. Note that in this case you do not connect black to red at any time. Just add a new rule to your list. In three-wire cable wiring, always connect black to black, red to red, white to white, and ground to ground as well as to the boxes. (Note the difference between this section and the previous section in which three-wire cable was connected to two-wire cable. There connecting red to black was safe. Here, where only three-wire cable is used throughout the circuit, both the black and red wires are hot but they are different hots and must never be connected directly together.) As mentioned, you must break off the small tab between the two brass terminal screws to separate the two receptacles electrically; the tab on the silver-colored side must be left intact. For the white neutral wire

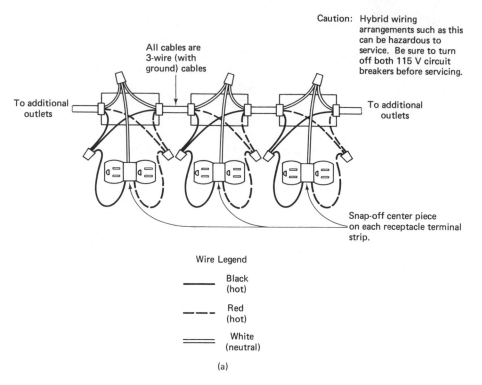

Caution: Hybrid wiring arrangements such as this can be hazardous to service. Be sure to turn off both 115 V circuit breakers before servicing.

All cables are 3-wire (with ground) cables

To additional outlets

To additional outlets

Snap-off center piece on each receptacle terminal strip.

Wire Legend

——— Black (hot)

— — — Red (hot)

═══ White (neutral)

(a)

Fig. 7-10(a). Using three-wire (with ground) cable you can have one receptacle of each outlet on the black 115-V circuit and the other on the red 115-V circuit. Ground connections are not shown here to prevent confusion.

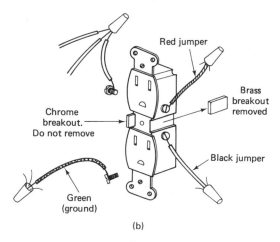

Red jumper

Brass breakout removed

Chrome breakout. Do not remove

Black jumper

Green (ground)

(b)

Fig. 7-10(b). Split receptacle wiring requires four wire nuts, one wire nut for the incoming and outgoing red wires plus a red jumper to one brass terminal; a second wire nut for black wires; a third for white neutral wires, and a fourth for ground wires and jumpers to the receptacle ground terminal and the box. The brass breakout must be removed to separate the black wire 115-V circuit from the red wire 115-V circuit.

connection, connect the two white wires with a short white jumper in a wire nut. Then connect the other end of the jumper to one of the screws on the chrome terminal. This provides a continuous neutral connection throughout the chain of outlets.

For the black wire, you will need a short black jumper wire to connect to the brass terminal. Gather the other end of the jumper with the incoming and outgoing black wires and fasten the three by wire nut. Do the same for the red wires. For the ground, make the usual connection of two short jumper wires to the box wall and receptacle ground terminal. Twist the other ends of the jumpers with the incoming and outgoing ground wires and fasten all four together with a wire nut. Repeat this wiring for each outlet in the circuit.

Caution: *In three-wire circuits never interconnect the black and red wires, or any wires spliced to them.*

WIRING ONE OUTLET ON ONE CIRCUIT, THE NEXT ON ANOTHER CIRCUIT

Another way of dividing your load would be to alternate boxes between circuits as shown on Figure 7-11. Again, using three-wire cable, you could run a chain of 20 boxes but still have a reasonable assurance that neither circuit will ever be overloaded. You do not have to change at alternate boxes either. You can group the outlets in any way you wish to divide the load, putting about half the outlets on each circuit.

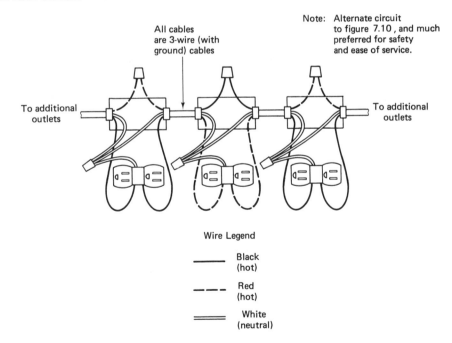

All cables are 3-wire (with ground) cables

Note: Alternate circuit to figure 7.10, and much preferred for safety and ease of service.

To additional outlets

To additional outlets

Wire Legend

—————— Black (hot)

— — — — Red (hot)

═══════ White (neutral)

Fig. 7-11. Using three-wire (with ground) cable you can alternate outlets on a branch to minimize chances of overload and blowing a fuse. Every other outlet is on the black 115 V circuit with bypassed outlets on the red 115 V circuit. Ground connections are made in the same way as shown in previous illustrations, but are not shown here in order to prevent confusion.

Note: *In each of the previously mentioned split circuits, you could install an ON–OFF switch in either the red or the black wire. Then with one switch you could turn a whole chain of outlets or any group of them on and off while the other outlets or other halves of the split receptacles are always live. For example, this will allow you to connect all lamps in a room to the switch. Then by turning the switch on or off you can control all lamps in the room but leave clocks, TV, or air conditioner in operation. Switch controls are described in the next chapter.*

THREE-WIRE CIRCUITS FOR HEAVY LOADS

Wiring for 230 V or for 115/230-V combinations can take many forms depending on the appliance (electric range to clothes dryer) that you wish to hookup. Generally, though, these circuits will take heavier wires than normal 115-V lighting circuits; common sizes are No. 10, 8, or 6 because current loads are very heavy.

Ranges and dryers are often grounded through the neutral (white) wire in the cable, but your local code could require some other arrangement. Check the code and power company for details first before you do any 230-V work. Some appliances using 230 V but needing a ground can be grounded through the white (neutral) wire. Check your code.

WIRING BOTH 115 V and 230 V RECEPTACLES IN A DUPLEX OUTLET

If you want both 115 V and 230 V to be available at the same outlet, you can also do it with a three-wire cable.* Buy a special duplex receptacle; remember that the blade arrangement for 115- and 230-V plugs is different so neither can be plugged into the wrong receptacle.

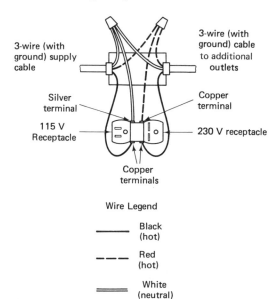

Fig. 7-12. Wiring a special duplex outlet with 115V on one receptacle and 230V on the other. This duplex receptacle is supplied by three-wire (with ground) cable and can be added to any position on a three-wire (with ground) branch circuit. Ground connections are not shown.

*These should not be used in dwelling-type units.

Besides different blade arrangements in the two receptacles, the duplex unit has a special arrangement for its terminal screws. Usually this means that one silver-colored and one brass terminal (separated) are on one side of the unit (see Fig. 7-12) and two connected brass terminals are on the other, the extra brass terminal being associated with the 230-V receptacle. Now when connections are made, two hot wires, one red and one black, can be connected across this receptacle to give 230 V. The other receptacle takes a standard wiring arrangement allowing the black wire to feed through to the next outlet as usual. The red and white wires must be spliced with wire nuts as shown.

Duplex outlets can be used as isolated outlets in a three-wire circuit, or there can be an entire chain of them, as you wish. Just be careful not to mix up the wires or jumpers—black to black, white to white, and red to red is correct. Also, because 230-V circuits often are used for heavy-current loads, this usage may require the use of a heavier three-wire cable throughout the circuit. Check this point carefully.

WIRING A 230-V RECEPTACLE FOR A SINGLE LARGE APPLIANCE

The last major three-wire circuit to consider is the straight-forward wiring of a 230-V receptacle for a major appliance. You will probably have only a single outlet drawing heavy current to an appliance such as a clothes dryer. In this case, the current-carrying capacity of the receptacle and the plug becomes very important.

There are many varieties of plug and receptacle blade arrangements, depending on current and voltage requirements. Your best bet might be to make a sketch of the plug-blade arrangement of the appliance you have or intend to buy and take the sketch with you when you shop for the receptacle. See Figure 7-13 for a typical electric range or clothes dryer hookup. Connect the three-wire cable of proper size for the current load to the box (use a three-wire cable without ground wire), then fasten the black and red wires to the two brass terminals of the receptacle to have 230 V available across the receptacle. Connect the neutral wire of the cable to the receptacle and to the box as a ground (bonding) wire. It may be used as a grounding lead in accordance with NEC requirements and can still serve as the neutral wire for both the red and black hot leads.

INSTALLING A GROUND-FAULT CIRCUIT INTERRUPTER
FEED-THRU RECEPTACLE

If you want the protection of a GFCI on a circuit and do not have a circuit breaker with this feature, you can install an Interrupter Feed-Thru Receptacle at any point of the circuit beyond the overcurrent protection device. Many types of interrupters are commercially available and they are usually supplied with wiring instructions when you purchase them.

We will take a typical unit (a Leviton Mfg. Co. unit) for illustration. Install this type of device only on a grounded a-c circuit protected by a fuse or circuit breaker. When installed as the first outlet of the circuit, it will protect every outlet

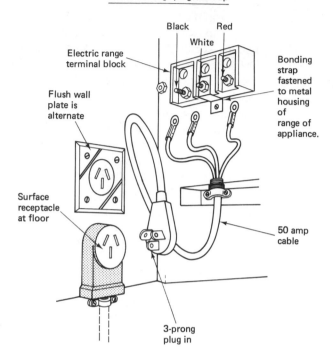

Fig. 7-13. A 240-volt hookup to range or other major appliance.

on the same branch circuit. If it is installed further down the chain, all outlets between the interrupter and the overcurrent device will not be ground-fault protected.

To install this device as a replacement for an existing non-GFCI dual receptacle or as the first receptacle of a new branch circuit, turn off power to the circuit and proceed as shown in Figure 7-14, using a 3½ in. deep box.

Suppose that you are installing this device as a replacement. Several minor complications may arise. You will probably want it as the first outlet in the circuit. Unless you can actually trace the wires to the service entrance, you will have to do some trial and error testing with the fuse in and out and a test lamp to see which outlets are on the circuit. Then disconnection of a few receptacles and some guesswork will allow you to determine which outlet is the first one. Your troubles are not over yet. Because the interrupter is a polarized (one way) device, you still have to know which is the incoming hot wire from the service entrance and which the outgoing hot wire to the remaining outlets. You can do this best by turning off circuit power and then disconnecting one black wire from the receptacle. Carefully tape or cover the end of the wire and turn circuit power on. Plug a lamp into the outlet and turn it on. If it lights, your service entrance wire is the one connected to the receptacle. If it does not light, the other wire is the service

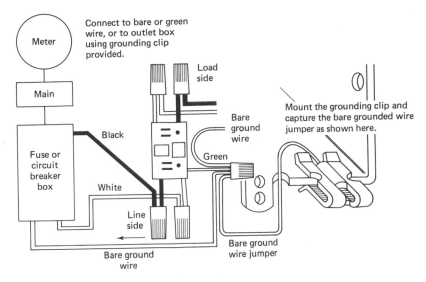

Fig. 7-14. Installing an Interrupter Feed-Thru receptacle (Courtesy of Leviton Mfg. Co.)

entrance hot line. In either case, turn off power again, identify all wires with tape or tags of some sort, then disconnect wires and remove the receptacle.

The procedure for connecting the GFCI is the same as for replacement or for a new installation. You must keep track of the incoming (service entrance) and outgoing (outlet chain) wires. No black or white jumpers are required because the GFCI has short lengths of wire for attachment. Two wires are black, two white, and the ground wire is green. Your incoming black and white wires are attached by wire nuts to the black and white wires of the interrupter, which are marked "LINE." Similarly, outgoing wires are connected to the GFCI wires marked "LOAD." Connect the interrupter green ground wire, the incoming and outgoing bare ground wires, and a ground wire jumper, green or bare (four wires in all), by a wire nut and connect the other end of the ground jumper to the box by a screw or grounding clip.

Turn on circuit power. On this unit the power is on and the circuit is working for your protection when the red "R" (reset) button is pushed in and remains in and a white band is not visible in the reset window. A yellow test button allows for recommended periodic test of the GFCI. Press the yellow "T" (test) button and the red button will pop out, exposing a white band. This simulates what will happen if a ground fault actually occurs and turns power to all the outlets off. To restore power, push the "R" button back in. If the white band does not show when the test is made, recheck to be sure the installation was correctly made. In service, the interrupter is used like any other grounded receptacle while providing protection for itself and all subsequent receptacles on the line. Other manufacturer's interrupters will work in much the same way as this one, though installation instructions may vary slightly.

8

SWITCH CONTROLS

Circuit Interruption – Switch Loop – Split Circuits – Three-Way Switches

Much of your system wiring requirements are covered by the circuit discussions in Chapter 7. However, there are many circuits in which you will want switch controls—wall switches to control a ceiling light, lights turned on and off by switches at top and bottom of stairs, garage and porch lights, or a whole string of outlets. The specifics of each are given in the following sections.

SWITCH CONTROLS WITH TWO-WIRE CABLE

You have the fundamentals. Now we can discuss electrical control of the receptacles or of light fixtures by means of ON—OFF switches. This can mean anything from turning an entire branch circuit on and off by means of a switch to controlling only one receptacle or fixture of a circuit. We will cover the most likely possibilities using two-wire cable, then discuss control circuit possibilities with three-wire cable.

Switch in Hot Lead vs. Switch Loop

Suppose you have a ceiling light fixture on a branch circuit. How do you manage ON—OFF control of the light? You could wire it in permanently just as you do a dual receptacle and use an ON—OFF pull chain attached to the fixture. This, though, is not as neat or as convenient as a wall switch. You might easily decide that a wall switch is what you want, so now you have to decide how to wire that switch.

Three probable configurations will occur:

1. If the light fixture is the last device on the branch circuit, you can lead your cable through a switch box at shoulder height near the door of the room and connect the switch in series with the black lead. All outlets prior to the switch are unaffected, but as you turn the switch on or off you control the lights of the ceiling fixture independently of the outlets.

2. If the light fixture is the last active component of a circuit but you cannot conveniently get to the input black lead to interrupt it with a switch, you might consider extending the fixture cable beyond the fixture and down a nearby wall to a wall switch. This control is just as effective as a switch in the fixture input, but it does entail some complication, which we describe below. This configuration is called a *switch loop.*

3. A slightly more complicated situation is if you have a chain of outlets and decide to break out a side chain through a switch to ceiling fixture. A sideways switch loop will do this job.

Let's discuss each of these control situations. None are really difficult but each is different in some small way.

Controlling a light fixture from a switch ahead of the fixture. This situation is shown in Fig. 8-1, using regular two-wire with ground (NM 14-2 plug ground) cable. All wiring ahead of the switch box could be any of the two-wire circuits previously covered, such as a chain of dual receptacle outlets that are always on. Note that in the switch box, the white (neutral) wire is simply spliced through as though the switch box was not even there. The black wire, however, is connected to two brass terminals at each side of the switch housing. As the switch is opened

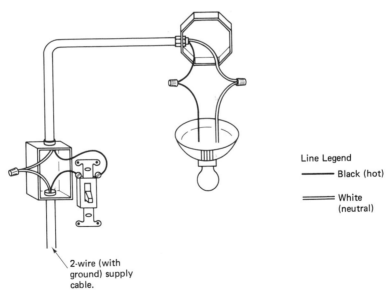

Line Legend

———— Black (hot)

===== White (neutral)

2-wire (with ground) supply cable.

Fig. 8-1. Nonmetallic cable is here wired from switch to fixture. Straps, not staples, are used in mounting this type of cable, and special connectors are used to attach cable to boxes.

and closed the light alternately goes off and on. The switch therefore only interrupts the current in the hot lead, as we have said it must. Ground connections for switches are just slightly different than those for dual receptacles if the switches do not include a ground terminal. At each switch box, the incoming and outgoing ground wires must only be connected together and to a ground jumper with a wire nut (three wires, not four, as we have used for receptacles). The jumper is then connected by a screw or ground clip to the box or by means of the screws that fasten a metal strap of the switch structure to the box.

Controlling a light fixture from a switch beyond the fixture (switch loop). This is a job that can be done as part of an original installation or as a modernizing add-on to a pull-chain ceiling fixture. The procedure is virtually the same for both jobs, so let's just assume one and start from there. Suppose you have a ceiling fixture with pull chain and want to install a chandelier. How do you modernize the fixture for wall switch control?

Figure 8-2 shows the procedure for adding a switch-loop circuit to a light fixture. Shut off power to the branch at the service entrance panel. Disconnect the black and the white wires from the fixture and the ground wire from the outlet box. Install a new NM 14-2 with ground wire cable from a switch box on the wall to the outlet box, as shown in Fig. 8-2. Connect the two ground wires together under one screwhead fastened to the box or group the ground wires with a jumper in a wire nut and connect the jumper to the box. Connect the incoming black wire to the switch cable black wire with a wire nut as shown.

For the fixture, connect the existing white neutral wire by way of a wire nut and white jumper to one terminal of the fixture. Now are you ready to violate a

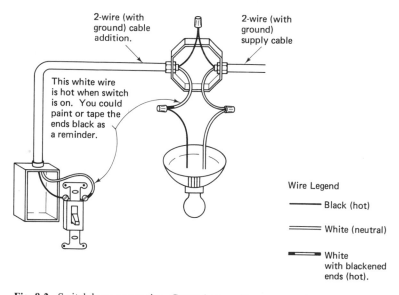

Fig. 8-2. Switch loop connection. Ground connections are not shown.

standing rule about color codes? *Connect the fixture black wire by wire nut to the white wire of your switch cable.* That's right, we connected white to black, forbidden by the rule. This is another of life's little exceptions. You see, what we have done is made the 14-2 wire into a continuous loop, closed through the switch. Look at the circuit in Fig. 8-2. From the incoming black wire through the switch, *all* the wires are hot. The switch does not put any load on the wire, so it is still hot for the white wire that loops all the way back to the fixture. Only *after* the fixture is there any contact to neutral.

But why did we use white for a hot wire? It was out of necessity. If you could buy a standard length of cable with two black wires, this would be the circumstance to use it. But to the best of our knowledge, it is not available. Better still, if you are working with conduit, you can use two black wires together. Then you would not have to make an exception to the rule. The NEC used to require that both ends of the white wire be painted black as a reminder of the fact that it was acting as a black wire. This is not true in the present NEC, but some local codes may still include this rule. You'd better check your local code. However, you could label the wire for your own sake to remind you that the white wire used in a switch loop is *hot.*

Remember to connect the ground wire to the switch outlet box by a screw in the wall or clip on the box edge.

SWITCH CONTROLS WITH THREE-WIRE CABLE

As you might expect by now, three-wire cables offer much greater versatility for control circuits. From a two-wire circuit, a three-wire extension or side branch will allow many special control circuit arrangements. In this section we discuss three typical types that exhibit most of the principles involved. From then on you can probably figure out other arrangements if the need arises in your layout.

Wiring a Switch to Two Fixtures—One Pull-Chain Controlled

Refer to Fig. 8-3. From the two-wire with ground incoming cable, connect to a three-wire with ground outgoing cable to the fixtures. Ground connections are made as normal. The only special feature here, or complication if you'd prefer, is that the three-wire cable allows you to feed one fixture by the black wire of the cable without going through the wall switch and the other fixture from the red wire of the cable. This is somewhat like feeding alternate outlets, as we discussed in the last section. The red wire is therefore only hot when the switch is on; the fixture supplied from the black wire may have its own pull-chain ON—OFF switch. The only place where our color-code rules are violated is at the first fixture, where a red wire must connect to a black wire of the fixture. Note that this circuit shifts from two-wire cable to three-wire cable and back to two-wire cable between the two fixtures. It should be a simple enough connection, though, if you keep track of which wire is which.

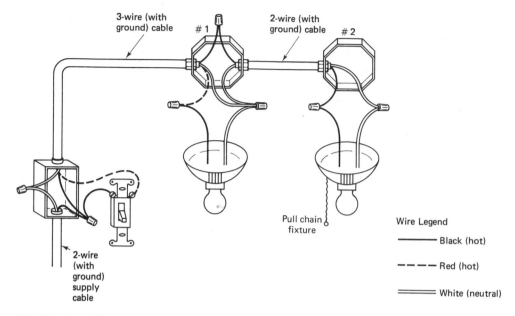

Fig. 8-3. Controlling one fixture (#1) by wall switch while leaving #2 always on. Fixture #2 may be a pull chain fixture as shown, clock, or convenience outlet. Ground connections are not shown.

Adding a Switch and Outlet to an Existing Fixture

If you have a ceiling fixture supplied by a two-wire cable, you can add an outlet and a switch with the circuit shown in Fig. 8-4. Grounds are again connected as usual, using wire nuts and jumpers to the boxes. Disconnect the black and white incoming wires to the fixture and connect a three-wire cable to the incoming wires and fixture, as shown. Again, you are connecting a red wire to a black wire, but it cannot be helped. At the switch box, the black wire from the three-wire cable and the black wire from a new two-wire cable connect to one brass terminal of the switch; the red wire goes to the other brass terminal. The white wire of the three-wire cable bypasses the switch and connects to the outlet. A black wire from the outlet is then connected back to the same brass terminal of the switch, to which the incoming black wire connects. If the switch and outlet are both mounted on the same box, and an additional two-wire cable is not required.

To check the operation of this circuit, note that the black wire of the incoming two-wire cable supplies power to everything. At first it bypasses the fixture, then bypasses the switch; the outlet will be live at all times because it always has a black wire connection and a connection back to the source via a neutral wire that bypasses the switch and ceiling fixture. Now when the switch is turned on, current will flow back through the red wire, through the fixture, and to neutral so that the ceiling light goes on.

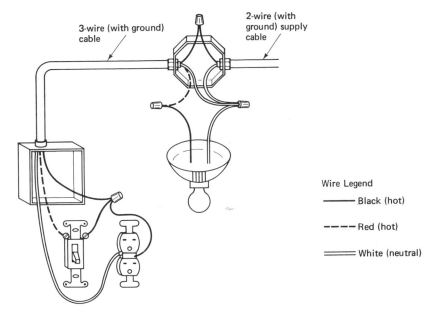

3-wire (with ground) cable

2-wire (with ground) supply cable

Wire Legend

———— Black (hot)

– – – – Red (hot)

═══ White (neutral)

Fig. 8-4. Circuit for adding an outlet and switch to existing fixture.

Adding a Second Fixture and Two Individual Fixture Switches

Figure 8-5 shows a more complicated arrangement still: A two-wire feed cable to a ceiling fixture connects to a three-wire cable, a second fixture, a second three-wire cable, and dual switches. With ground wires connected as usual, other connections are made as shown. Two color-code discrepancies occur; a red wire is connected to black in the first fixture and white to black in the second. It would be good to paint both ends of the offending white wire to be sure you remember it is being used as a hot wire.

Let's check the circuit. Again, the black incoming wire is the key. It bypasses both fixtures on the way to the switches. Then it is connected to both switches. When the right-hand switch is turned on, the right-hand fixture in our illustration goes on; note that it has a neutral connection back to the incoming cable. The same is true for the left-hand switch and left-hand fixture. Either fixture may be turned on independently.

Switch-Control of Light From Two Locations

Now we come to some complicated but very useful circuits; control of a light fixture from two different locations. These circuits require a different type of switch than we have previously discussed. This switch is called a three-way switch and has three terminals instead of the usual two. One terminal, colored darker than the other two and mounted on one end of the housing, is called the common

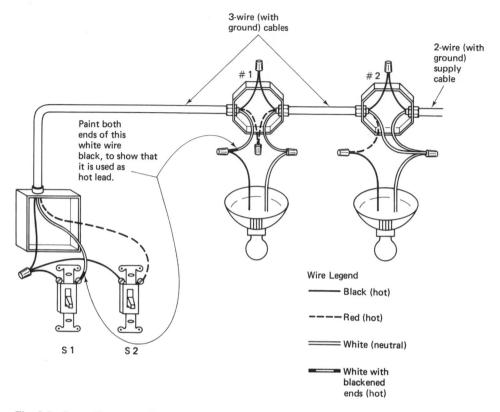

Fig. 8-5. Controlling two ceiling fixtures from two separate switches requires two three-wire (with ground) cable as shown above. You should be able to trace how S1 controls fixture #1, and S2, #2. Note that incoming black lead passes right through the fixtures to the switch. Ground connections are not shown.

and is always live; connect this to the black wire in all cases. When you buy a three-way switch, check your supplier to be sure you know which terminal is the common terminal. The two terminals mounted at the other end of the housing alternate in function. When the switch lever is in one position (for example, pressed up) one terminal is connected to the hot wire and the other is not; with the lever pressed down, the terminals change state.

That is how each switch works, but to explain the operation of the circuit we will have to resort to a schematic diagram before moving on to the wiring. Fig. 8-6a and b shows the electrical hookup for this type of circuit. As you can see, switches S1 and S2 are hooked up in the opposite way from each other. This provides the special feature. With the switches set as shown, if either switch is thrown to the opposite position, the light will go out. If S1 is thrown, the schematic will look like that shown in Fig. 8-6b. There is no continuity to the fixture and the light is off. Now, however, if either switch is thrown, say S1 back to where it was or S2 down, the light goes back on.

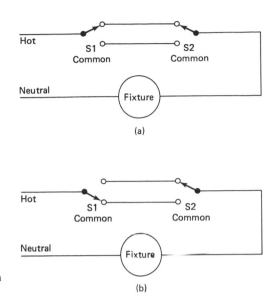

Fig. 8-6. Electrical hookup for switch control from two locations.

(a)

(b)

Figures 8-7 and 8-8 show two ways to wire this type of circuit, either with the switches beyond the fixture (Fig. 8-7) or with the fixture positioned between the two switches (Fig. 8-8). In either case, both circuits are electrically identical to the schematic diagrams shown previously. There are many variations of this same circuit, but you might think of using the switches ahead of the fixture or even adding outlets beyond the fixture; this last variation is shown in Fig. 8-9, but is not described in our discussion.

Connecting Three-Way Switches beyond the Fixture

Wiring these circuits may look complicated, but we will go through fairly slowly. Let's discuss Fig. 8-7 first. The grounds are connected as usual. Note that both the incoming cable to the fixture and the outgoing cable to the first switch are two-wire cables with ground. From the first to the second switch, a three-wire cable is required.

You can probably determine which cable is which on the schematic (Fig. 8-7). At the fixture, connect the incoming and outgoing black wires together. Connect the fixture white wire to incoming white wire and the fixture black wire to the outgoing white wire (you should mark this wire "black" to keep track). At Switch 1, connect the real black wire to the switch common and the white serving as black to the black wire of the three-wire cable that is the next link (mark the white wire black). The other two wires of the three-wire cable, namely the red and white, are connected to the noncommon terminals of the switch. (Each wire can be connected to either terminal). Connect the other ends of these wires to the same terminals on switch 2 (again, the connections can be made to either terminal). This white wire is also being used as a black wire so mark both ends black. Now connect the black wire of the three-wire cable to the common switch 2 and

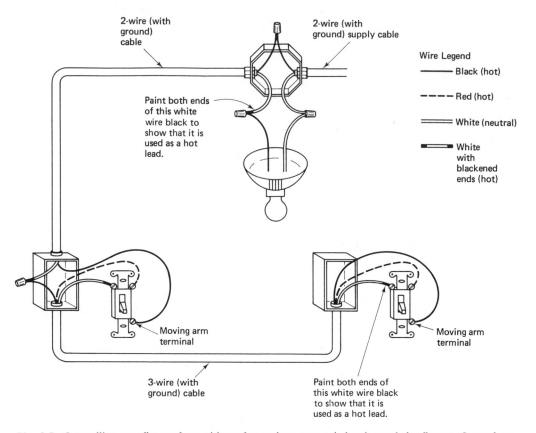

Fig. 8-7. Controlling one fixture from either of two three-way switches beyond the fixture. Ground connections are not shown.

remount all switches and the fixture into the boxes. Check that, if the light is on, changing either switch lever will put it out; if the light is out, either switch will put it on.

Wiring a Fixture Between Two Three-Way Switches

If the fixture is electrically between the two switches, then the supply must come into the fixture in a two-wire cable and split to two sides of the fixture by way of three-wire cables as shown in Fig. 8-9. A fast trace of this circuit will show that the hot line (black) comes into the fixture and is spliced through to a black wire of one three-wire cable (which is the right one in our illustration). This black goes to the common of the right-hand switch. The red and white (acting as black) wires run to the other two terminals of the switch and back to the fixture box, where they are spliced through to red and white wires of the left-hand cable. These in turn go to the noncommon leads of the left-hand switch and, if both switches are properly set electricity, leaves through the common of the left-hand switch back to the fixture. Because the fixture neutral is permanently connected,

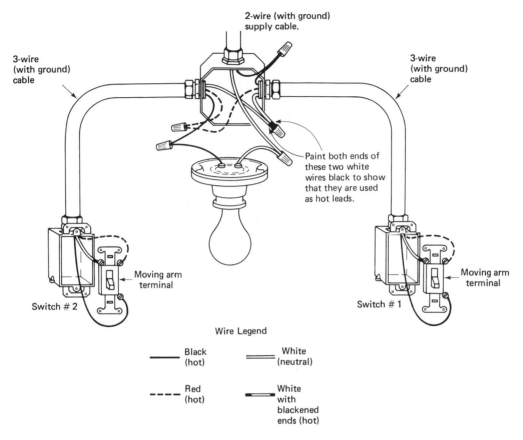

2-wire (with ground) supply cable.

3-wire (with ground) cable

3-wire (with ground) cable

Paint both ends of these two white wires black to show that they are used as hot leads.

Moving arm terminal

Moving arm terminal

Switch # 2

Switch # 1

Wire Legend

——— Black (hot)

═══ White (neutral)

- - - - Red (hot)

━━● White with blackened ends (hot)

Fig. 8-8. Controlling a fixture from either of two three-way switches to either side of the fixture. Ground connections are not shown.

the light goes on. Flipping either switch interrupts the circuit and the light goes out. Electrically, this circuit is identical in function to the schematics shown in Figs. 8-6a and 8-6b.

Multiswitch Control of a Fixture

Less common than the three-way circuit just discussed, circuits using four-way switches allow control of a single fixture from any number of locations. If the light is on, changing position of any switch lever will turn it off and vice versa. Like the circuits previously discussed, the fixture can be at any location in this type of circuit. All that is required is a three-way switch at each end of the chain, plus any number of four-way switches in between. The four-way switch has four terminals. The operation of the switch lever is shown in Fig. 8-10. In one position of the lever (Fig. 8-10a), terminals 1 to 3 and 2 to 4 are connected; in the other position (Fig. 8-10b), they cross and 1 to 4 and 2 to 3 are connected.

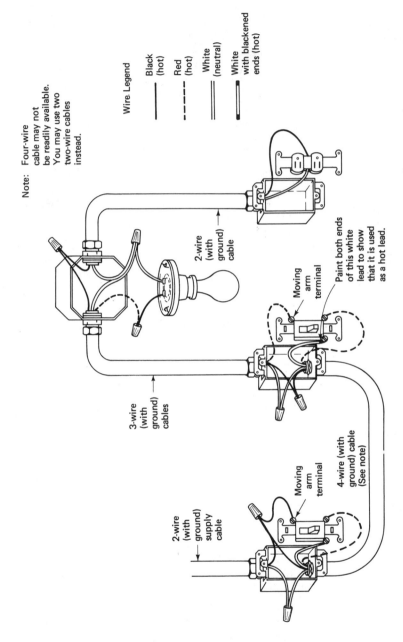

Note: Four-wire cable may not be readily available. You may use two two-wire cables instead.

Wire Legend

———— Black (hot)

– – – – Red (hot)

══════ White (neutral)

▬▬▬▬ White with blackened ends (hot)

2-wire (with ground) cable

Moving arm terminal

Paint both ends of this white lead to show that it is used as a hot lead.

3-wire (with ground) cables

Moving arm terminal

4-wire (with ground) cable (See note)

2-wire (with ground) supply cable

Fig. 8-9. Adding an outlet beyond a fixture controlled by two three-way switches. Ground connections are not shown.

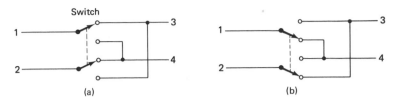

Fig. 8-10. Circuits for four-way switches.

Now consider putting one of these between two three-way switches as in Fig. 8-11. In this simplified, four-way switch representation, think of the switch crossing and uncrossing connections each time the switch lever is thrown.

In the position shown (S2 in Fig. 8-10b position), the light will be on, but changing position of any switch will turn the light off. From that position, changing any switch will put it on again. It does not matter how many four-way switches you put in the circuit, operation will always be the same as long as the chain begins and ends with three-way switches. Wire the switches with three-wire cable as shown in Fig. 8-12. Again, ground wires are connected as before, and any time you use a white wire for a hot lead, it should be marked black at the ends to remind you of that fact.

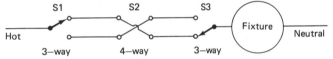

Fig. 8-11. A four-way switch between two three-way switches.

SPECIAL-PURPOSE SWITCHES

We could not possibly describe all types of switches that you will find on the market today. If we could, the list would be out of date tomorrow. However, there is a certain number of special-purpose units that can be discussed with the thought of expanding your knowledge of what can be accomplished. If you buy those or still more versatile types, you will probably be able to handle them by referring to what we have discussed so far and the normal instructions you will receive with the purchased product.

Silent switches, mercury or otherwise, fall into this class, but these are functionally identical to the toggle switches already discussed, except for the fact that they do not snap from one position to the other with a loud click. Instead, these operate silently, which is a boon in many places, such as a nursery.

Among the many other special-purpose switches, our favorites are as follow:

Dimmer Switches

Incandescent light fixtures can easily be controlled by a dimmer switch (in theater fashion) in place of a toggle switch. A dimmer switch fits in the same box and connects in the same way as a single-pole toggle switch. Most dimmers on the

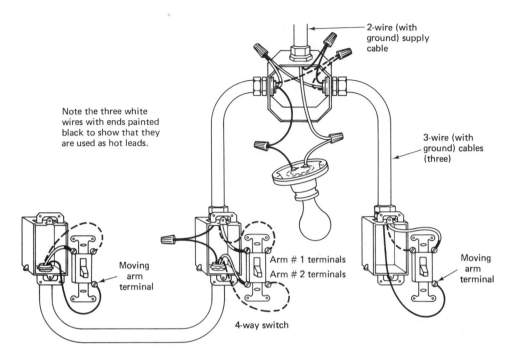

2-wire (with ground) supply cable

Note the three white wires with ends painted black to show that they are used as hot leads.

3-wire (with ground) cables (three)

Moving arm terminal

Arm # 1 terminals

Arm # 2 terminals

Moving arm terminal

4-way switch

Fig. 8-12. Installation of one four-way switch into a circuit with a fixture controlled by two three-way switches. With this circuit the fixture may be controlled from any one of the three points. Any number of four-way switches may be added to the circuit. Ground connections are not shown. Note the three white leads with ends blackened to show that they are used as hot leads.

market have a round knob that you push in both for turn-on or turn-off and rotate for dimming control. By dimming the lights, they also reduce wattage and so help to reduce electric bills. Dimmers work by automatically cutting off part of each electrical cycle, thereby reducing the time in each cycle during which the bulb filaments are heated. Typically, a 10% reduction in ON time per cycle saves 15% in wattage; 25% saves 36% in wattage, and so forth. Besides these savings, bulb life can be extended by 4 to 40 times, respectively. A lamp controlled by a dimmer loses some brightness, even in the full ON position; it will never be exactly as bright as a lamp operated by a simple ON–OFF switch, but the difference is slight. In place of screw terminals or push-in terminals, dimmer switches often have short wire leads extending from them. Connect these to the original switch wiring with wire nuts and fasten the dimmer onto the box with cover in place over the dimmer. Be sure that the dimmer has a high enough capacity for the wattage of the fixture it will control. Typical dimmer capacity is 600 W, which will cover a fair-sized fixture such as a dining-room multilamp chandelier. You should not consider using dimmers for anything but fixtures that use incandescent bulbs. Fluorescent lights require special devices, so normal dimmer switches should definitely not be used for them.

Lighted Handle Switches

A choice of switches is available in this category. One type has a handle that glows (with a miniature neon lamp) when the switch is off. This makes the handle easy to find in a darkened room. This switch has two terminals, like a toggle switch, and can easily be used for a replacement on a one-to-one basis. As an added feature, if this switch operates a remote light, such as an attic or garage, it tells you if the light is out.

Another type of lighted switch, the pilot light switch, has a handle that glows when the switch is on. This is designed as a reminder that a remote light, not visible from the switch location, is still on. This type has three terminals, two brass for the usual hot connection (black wire) and one chrome for the neutral (white) wire. This switch can only be installed if both the black and white wires enter and leave the box as in Fig. 8-1.

Timer Switches

Timer switches are of three basic types. A *delay toggle switch* will take a certain time, such as half a minute after being operated, before it snaps. This type can give you a slight delay after switching to give you, for example, enough time to get into the house from the garage.

A *timer switch* provides the same delayed shut-off, but for longer periods. You can use this type for a bathroom fan or similar device that you want to run for a time before turning off automatically.

A *time clock switch* can be set to turn lights or appliances on or off at a specific time on the clock (but remember that a power outage will throw these units out of synchronism with your preset schedule). These switches are connected differently from other switches that we have discussed. Both wires of the incoming and outgoing cables are connected to these switches, as are both wires of the outgoing cable.

As applicable, some of the switch devices just mentioned are discussed again in our next chapter on lighting. More often than not, switches are used to control lights for utility or for special effects.

9

LIGHTING

Lighting Effects - Incandescent And Fluorescent Lamp Installation

LIGHTING EFFECTS—PLAIN AND FANCY

Next to motors, which at times demand almost 50% of the household electrical power, lights are by far the most important consumers of electricity. Designing your home lighting environment is therefore a matter for careful consideration.

In planning an entire lighting system, or simply replacement of an outmoded fixture, try to keep in mind lighting efficiency as well as decorative effects. Wall and ceiling fixtures, particularly, exert a profound influence on each room. Many older fixture and chandelier forms that were long abandoned have again gained popularity. These fixtures are also available with new, modern features that enhance their value—anything from a dim, soft glow to bright, shadowless illumination can be provided by a touch or at most a twist of a switch. Do you have a fixture of real beauty or value that you would hate to give up? Why not modernize it by rewiring it for greater safety and enhanced performance. Remember that the heat generated by years of use could have loosened sockets, corroded contacts, or dried fixture wiring so that the insulation is cracked and unsafe. Replacement parts are easy to get and rewiring can help you restore a charming older fixture.

Ceiling fixtures are generally used for general room illumination. When well chosen, they can create a special atmosphere that will determine the entire character of a room. Fixtures can also be combined with structural features of the room (alcoves, dining ells, etc.) for very pleasing effects. Shadowless light can be achieved in working areas such as the kitchen by using several fixtures widely spaced to flood the area from various angles. The principle here is simple enough. A shadow cast by light from one direction is eliminated by light flooding the shadowed area from a different angle. This effect is especially easy to achieve with light-colored walls and several large fluorescent fixtures.

Other fixtures offer various special features, such as high-low or three-way switches, to provide a range of illumination from romantic candlelight to maximum illumination. When selecting a fixture, just be sure that the manufacturer's wattage rating and fixture size will be large enough to produce the amount of light you will need for the size of the area. Often, more than one fixture is needed. For example, a large kitchen may need a pair of two-tube fluorescent fixtures to cover the entire area and to provide an even distribution of light throughout the room. Remember, good lighting does not mean just enough light; it also means control of glare, which can make reading or sewing tedious.

Because modern ceilings are so low, many people think that chandeliers cannot be used, except possibly over a dining table where no one could bump their heads. This is not the case. New fixtures have been designed in modified versions that retain traditional design but provide the maximum possible headroom in everyday use, with some fitting snugly against the ceiling.

Before buying any fixture, check your choice against the following list. Keep these points carefully in mind while shopping.

> Incandescent lamps should never be closer than ¼ in. to the enclosing globes or diffusion shields.
>
> Bare lamps should never be visible from the normal viewing angle.
>
> Top or side ventilation is almost essential in a fixture to lower operating temperatures and extend bulb life.
>
> Inside surfaces of the lamp or shades should be of polished material or bright enamel coated to reflect light and spread it through the room.
>
> The fixture's shape should be designed to efficiently diffuse light. When buying a fixture, always insist that it be turned on in the store to check for the spread and distribution you want.
>
> The best material for enclosures and fixture shades is plain or textured glass or easily cleaned plastic.

LIGHTING CONTROLS

Besides selection of the fixtures, you must also consider the control effects that you want. Dimmers and timers, which we mentioned in the last chapter, come first to mind.

Light dimmers allow you to adjust the room lighting levels to match your moods and activities at the twist of a knob. Also, they save you money by dramatically cutting back on the use of electricity and increasing light bulb life. Though fluorescents require special devices, almost any incandescent light fixture can be controlled by an ordinary solid-state dimmer control of appropriate wattage rating.

The heart of the dimmer control is a semiconductor device called a *triac*, an electrical component that is similar to the transistor. The triac acts essentially as a rapid-fire ON—OFF switch that dims bulb brightness by controlling the amount of current that can flow into the bulb during each electrical cycle. The triac is wired in series with the bulb and the power line. During operation it repeatedly turns on and off, in step with the current cycles, to allow current to flow for only a portion of each cycle. You cannot see this effect because the switching occurs so rapidly. With current flowing for less and less of the full cycle as you turn down the control, the bulb glows less brightly because its filament is not heated to as high a temperature. You therefore save electricity use and save wear on the bulb filament, which extends the bulb life. Even a small reduction in operating temperature can extend bulb life by many months or even years longer than you would normally expect. There is one drawback, though: Some dimmers create annoying radio interference you can hear on a nearby AM radio. They also interfere with wireless intercoms. Top-line models are the most expensive but are equipped with interference-suppression circuitry.

You could also consider wiring a timer into at least one branch circuit so that you can have lights go on and off at preset times whether you are at home or not. This is considered to be an effective security measure to deter intruders. A house that stays dark for several nights can often become a target for unlawful entry by announcing that the owners are away.

A timer is generally wired into the branch circuit near the service entrance so that it may open and close an entire circuit in accordance with your selected schedule. Lamps or radios plugged into the circuit and left turned-on go on and off as the timer cycles. The timer is essentially no more than an ON—OFF switch with a clock motor controlled by a 24-hr dial with ON and OFF setting levers.

Whether you use indirect lighting or a ceiling fixture, dimmer or time control, the illumination will be derived from either of two types of lamps, incandescent or fluorescent.

INCANDESCENT VERSUS FLUORESCENT LIGHTS

Incandescent lamps of the type invented by Thomas Edison consist of a glowing wire inside a glass bulb filled with inert gas. Fluorescent lamps provide light by electrical current excitation of gas inside a phosphor-lined tube. We discuss incandescent lamp installations first because they are simple in construction and are so familiar to everyone. We have discussed most of the wiring aspects of these installations, but a look at details of ceiling fixture hanging and wiring might be in order before going on to the more complicated fluorescent assemblies.

INCANDESCENT LIGHT CEILING FIXTURES

In Chapter 5, we covered the installation of ceiling outlet boxes by means of adjustable brackets or old work hangers. We would like to speak in more detail now on what to do with the wires within the fixture box, *with power off, of course.*

Assuming that the wiring for the fixture comes from a wall switch that you have just installed, you must be sure that the electricity is off. Turning the switch off is not enough. Someone could accidently turn it on just as you are touching the black and white wires, standing on a ladder. It is better to be sure that the entire circuit is off at the fuse (put the fuse in your pocket) or circuit breaker, then set the wall switch off for extra assurance.

In the same way that you connect cable to a dual receptacle outlet box, strip the outer and inner insulation from the cable and wires, insert the cable into the outlet box, and clamp the cable to the box wall. Remember that with an old work hanger, the box could possibly be shallow and the cable must enter and be clamped to ports in the top wall of the box rather than the sides. Connect the ground wire to the box by a screw or ground clip.

Fixtures come in a variety of configurations and are usually supplied with a set of instructions for installation and wiring. Let's look at the most common types. Figure 9-1 shows a mounting strap placed across the box front and fastened by a locknut to a threaded stud in the center of the box. If this is the type of mount your fixture requires, fasten the strap securely before proceeding. To connect the fixture wires, you could use three hands, one to hold the weight of the fixture and the other two for connection. A coat hanger is sometimes handy here. Twist it to shape and hook it to the fixture, then onto the fixture strap to hold the fixture while you work.

Strip both the house-wiring and the fixture-wiring ends (about ¾ in.) to bare metal. Your cable will have one black and one white wire for connection; the fixture will have two or more of these also. Twist the bare ends of all black wires together and connect them by a wire nut. Do the same with the white wires and the wiring is essentially complete. If, as sometimes happens, both fixture wires are the same color, you must check a bit further before connecting. Be sure to connect the black house wire to the fixture wire from the *center terminal* of the fixture bulb socket. The threaded socket portion (shell) should then be connected to the white (neutral) wire.

Close up the job by carefully tucking the wire terminations and wiring back into the box as you install the ceiling cover plate from which the fixture is

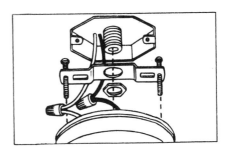

Fig. 9-1. To outlet box stud. (Courtesy of Sears, Roebuck and Co.)

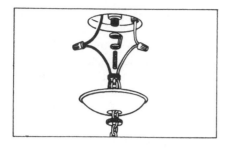

Fig. 9-2. To fasten a large drop fixture, simply screw hanger support onto threaded stud in the outlet box. Connect wires (Use solderless connectors), then raise canopy and anchor in position by means of a locknut. (Courtesy of Sears, Roebuck and Co.)

suspended. Fasten the fixture cover plate to the box mounting strap with the screws provided with the fixture. You will probably need an assistant to hold the fixture for a few minutes while you do this.

Variations on this installation procedure mainly involve the fixture suspension methods and techniques for fishing the fixture wires into the box. Sometimes, too, multilamp fixtures will have several black and several white wires to be connected. In this case, just use a large enough wire nut and fasten all blacks together and all whites together in separate wire nuts.

If your fixture is designed for attachment to a threaded stud rather than a mounting strap, you will probably need to lengthen the stud in the ceiling box. You can extend the stud by use of a pipe hickey and threaded nipple as shown in Fig. 9-2. Nipples come in various lengths so that the fixture can be suspended at just the right distance below the ceiling to accommodate the fixture cover plate. On some fixtures you will find that the fixture wires extend through the recess for the threaded nipple. If this is the case, simply extend the wire through the nipple and out into the box through the open side of the hickey. The wires can then be connected to your house wiring.

One last variety of mounting should be mentioned. Some boxes do not have a central threaded stud to which a mounting strap or pipe hickey can attach. For this type of box, fasten the mounting strap across the box front using screws; the straps are made with slotted mounting holes to accommodate different box dimensions. The fixture can then be fastened by screws to the mounting strap. If the fixture must be dropped further from the ceiling, a pipe nipple can be fastened by a locknut to the center of the mounting strap and the fixture mounted to the bottom end of the nipple. Again, wires may be led through the nipple for connection within the box if required. Fixtures may be mounted to wall-mounted boxes in practically the same way, making use of a mounting strap or pipe nipple, as appropriate.

FLUORESCENT LIGHTING

Fluorescent lights differ quite a bit from incandescents. A typical fluorescent lamp (see Fig. 9-3) is a straight glass cylinder. To conserve space, the cylinder is often formed into a circle. The lighting power of a fluorescent unit increases with length, typically ranging from 15 to 60 W for lengths of 18 to 60 in. Length for

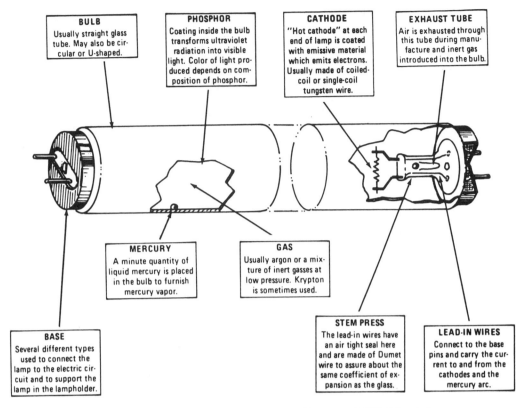

BULB
Usually straight glass tube. May also be circular or U-shaped.

PHOSPHOR
Coating inside the bulb transforms ultraviolet radiation into visible light. Color of light produced depends on composition of phosphor.

CATHODE
"Hot cathode" at each end of lamp is coated with emissive material which emits electrons. Usually made of coiled-coil or single-coil tungsten wire.

EXHAUST TUBE
Air is exhausted through this tube during manufacture and inert gas introduced into the bulb.

MERCURY
A minute quantity of liquid mercury is placed in the bulb to furnish mercury vapor.

GAS
Usually argon or a mixture of inert gasses at low pressure. Krypton is sometimes used.

BASE
Several different types used to connect the lamp to the electric circuit and to support the lamp in the lampholder.

STEM PRESS
The lead-in wires have an air tight seal here and are made of Dumet wire to assure about the same coefficient of expansion as the glass.

LEAD-IN WIRES
Connect to the base pins and carry the current to and from the cathodes and the mercury arc.

Fig. 9-3. Basic parts of a typical hot cathode fluorescent lamp. (Courtesy of GTE Sylvania.)

length, straight and circular units of the same cylinder diameter provide roughly the same light output. Fluorescent tubes last three to five times as long as incandescent bulbs in normal use and provide more light output (as much as two to four times as much) per watt of electric input. Not all types will operate at low temperatures, though, so buy the right type for unheated areas such as garages or outbuildings.

The color of the light produced by a fluorescent lamp depends on the phosphors which are used as the coating on the inside of the cylinder. While white light is the most prevalent fluorescent made, you can also obtain fluorescents that give off blue, orange, green, yellow, or pink light.

At each end of the fluorescent tube, metal electrodes, called *cathodes,* carry the current and perform several functions. First, they provide electrons. They are generally made with a tungsten wire filament covered with an oxide coating. The coating is a material that gives up electrons very easily. Electricity flowing through the filament heats it to make it glow, just like the filament in an incandescent lamp. The heat then drives electrons out of the oxide coating (boils them off, so to speak) and creates a cloud of electrons around each electrode. Second, voltage from your house electrical system is applied to the two electrodes so that at any given time one electrode is positive and the other negative, creating a voltage difference between the two ends. Under the electrical pressure of the voltage difference, the free electrons surrounding the electrodes suddenly migrate from one electrode to the other, rushing through the gas in the tube. This type of action is called an electrical discharge or arc and is similar to the discharge in an arc welder's torch or natural lighting.

The electrons flow back and forth between the electrodes as the line voltage alternates. Electrons streaming from one electrode to the other collide with atoms of mercury vapor inside and release ultraviolet light, which by itself is invisible. The ultraviolet light, in turn, strikes the phosphor surface on the inside of the glass tube and cause these atoms to fluoresce, thereby giving off visible light.

Rapid-Start and Instant-Start Fluorescents

Two more popular modern fluorescents are the instant-start and rapid-start units. The instant-start circuit applies a high voltage between the filaments to start the arc without the need of pre-heating. Instant-start ballasts are larger, heavier, and more expensive than pre-heat ballasts. A typical instant-start circuit schematic is shown in Fig. 9-4. Rapid-start units contain extra windings in the ballast transformer to allow continuous heating of the filaments; however, starting is not as fast as Instant-Start units. Some lamps have instant-start electrodes that do not require preheating. They do, however, require much higher starting voltages to shake the electrons loose without heating. The first pulse of voltage starts the electrons and the gas discharge then continues as in the starter-type lamps.

In the past 4 or 5 years, an improved fire-resistant ballast, called Class P, has been available and is now required for use in all new fluorescent fixtures. For replacement purposes, this will probably be the only ballast available. However, these units are deliberately more sensitive to heat. If, after you install one in an old fixture, you find that there is an excessive amount of nuisance tripping of the ballast shut-off, it will indicate that the old fixture runs too hot. Your only recourse then will be to install a new fixture.

Two auxiliary devices are required to make a preheat-type fluorescent lamp operate. One of these is a starter that acts as an automatic switch for the filament circuit, and the second is the ballast that performs several functions. The ballast acts first as a transformer to provide a momentary high voltage (a voltage surge) to get the arc discharge started, then acts later as a choke coil to limit current through the fluorescent tube. In a basic lamp circuit, the lamp filament, the ballast, and the starter are all connected in series. When the lamp is turned on, current flows through all the elements; the lamp filaments begin to glow slowly. After a second or two the contacts inside the starter snap open. This cuts off current to the filaments, leaving the gas in the lamp as the only path for current flow. Simultaneously, the sudden break in the circuit causes the ballast to perform its primary function, which is to develop a momentary high voltage surge

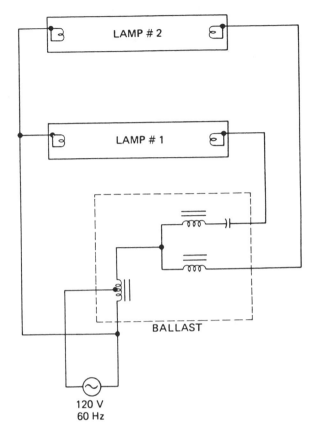

Fig. 9-4. Fluorescent fixture using instant start lamps. From Doyle, John M., *An Introduction to Electrical Wiring,* © 1975. Reprinted with permission of Reston Publishing Company, Inc., A Prentice-Hall Company, 11480 Sunset Hills Road, Reston, VA.

that is applied to the electrodes of the tube. This voltage acts on the electrons that were already loosened from the hot filaments, accelerates them to a high velocity, and causes a current to arc from one filament to the other, producing light as we described. At the instant the starter trips open, the filaments are disconnected from each other. One side of the a-c line remains connected only to one terminal of a filament on one end; the other side of the line remains connected to the ballast, which in turn connects to only one terminal of the other filament. The filaments themselves are no longer incandescent, but that is not important. Once the current flow has started by means of the arc through the gas, resistance drops to a very low level. Rapid-start and instant-start fluorescents do not use a starter.

Fluorescents can be installed quite easily if you purchase fixtures already internally wired. All you have to do is make black to black and white to white wiring connections from the outlet box to the fixture wires. Most fluorescent fixtures are made with screwholes and knockouts to fit standard outlet boxes. However, it is not sufficient to ground the outlet box if there is any danger of poor contact between it and the body of the fluorescent fixture. For safety, use a grounding jumper from the outlet box ground to the fixture housing. Remember, though, if your fixture contains any internal switches, they must be wired into the black line, never into the white (neutral) line. It goes without saying that any wall switches that have been wired in accordance with Chapter 8 must only interrupt the black wire.

10

SPECIAL CIRCUITS

Bells–Furnace–Security–Intercom–
Emergency Generator

Have you ever replaced a bell or buzzer, installed a chime assembly, or cleaned the contacts of your furnace thermostat? You could easily do this following a manufacturer's instructions without ever noticing that these devices are different from the other electric units in your house. You could in fact never have to worry about them because they can function indefinitely without replacement. But, if you do not investigate them, you may never discover the wonderful realm of low-voltage circuitry. Similarly, you could wire your entire system without thinking of installing other specialized circuits—security, intercoms, lightning arrest, emergency generators, and the like—all helpful aids that can make your home environment that much more enjoyable and safer. In this chapter we will explore these unique circuits, which in some ways are only the tip of the iceberg of possibilities that electricity can bring into your home.

LOW-VOLTAGE BELLS AND CHIMES

If you have assimilated what we have discussed up to this point, manufacturer's equipment installation instructions should be the only additional instruction that you need now, particularly in the area of low-voltage devices that operate typically at 6 to 24 V and are much less dangerous than 115/230-V circuits.

The most likely place you will use low voltage is your doorbell/buzzer/chime arrangement. You might want to change your bell or buzzer to chimes (most likely a dual system), with characteristic chime tones that distinguish between front door and back door visitors.

The instructions packed with the equipment you purchase will probably be clear enough. However, you may buy the transformer separate from the chimes and wiring and have to piece the instructions together. To aid this process, we describe the installation procedure for a typical dual-chime system in the following discussion.

First, let's give you some background. In years past, doorbells were usually operated by two or four D-cell batteries, providing a total of 3 to 6 V of dc. These systems are now old and inefficient. More recently, doorbells and buzzers were designed for operation from transformers that provided 6 V of ac from the household line. Although better than battery operation, these systems are also somewhat obsolete because they cannot be used to operate the newer buzzers or bells that operate on 10 V and chime sets that need 16 to 24 V.

Whatever the voltage is, you have to buy a transformer that supplies the amount required for your unit. Usually if you replace an old bell or buzzer with a chime unit, you will also have to install a new transformer. Figure 10-1 shows how to hook up a single doorbell so that it can be operated from three different places, front, back, and side door for example. However, this is not the ideal arrangement because you won't know at which door the visitor is. A better arrangement (see Fig. 10-2) shows the hookup that operates a front doorbell or chime from the front-door button and a buzzer from the back-door button.

The heart of your system will be the transformer. Its size will depend on the

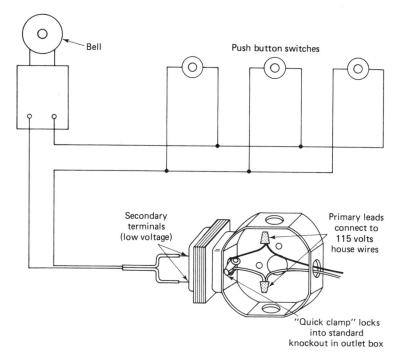

Fig. 10-1. Circuit for a single doorbell operated from three different locations.

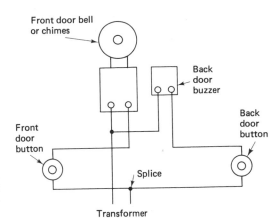

Fig. 10-2. With this circuit you can get a chime at the front door and a buzzer at the back door to tell where your visitor wants to enter.

volt/ampere capacity, but it is usually relatively small, probably about the size of your fist. Usually, permanently attached black and white wires extend from the unit. These are the leads to the transformer's primary winding and must be connected to a 115-V branch when you install the transformer.

Some transformers are made so that they can mount directly onto an outlet or junction box with primary wires running directly into the box. This might be your best choice because they are 115-V leads and all connections or splices must be made inside of a box just as with the rest of your regular house system. This is the only part of your low-voltage system installation that is dangerous. Whenever making this connection make sure that circuit power is off at the service entrance before proceeding. Use wire nuts to splice the leads to the house wiring at any convenient junction box.

Secondary wires of the transformer are the low-voltage leads. They are not attached to the transformer when you buy it because it is easier to make this connection directly from the wires that you connect to the bells, buzzer, or chimes. Two terminal screws are provided on the transformer to connect the secondary wiring. Because the voltage ranges from 6 to 24 V and there is no great shock danger, you do not need to use cable such as BX or NM. No. 18 bell wire, similar in appearance to lamp cord, is the most commonly used wiring and can run along baseboards or the exposed surfaces if the system is installed after the house is built. Insulated staples hold the wire in place quite easily and the wire is thin enough to be unnoticeable. In new work, of course, you can hide it within the walls.

To understand the circuit, consider the transformer as the source, the doorbell button as the switch, and the chime or bell as the load. Install your wiring to provide a single front-door button or front- and rear-door arrangement as shown in Figure 10-2.

Note that the No. 18 bell wire is available in single-, double-, or triple-wire form. Leads are not color-coded, so you will have to mark them yourself to keep your circuit organized. Wiring can be run as is unless a local code prohibits it. Be sure to protect the wiring from damage in particularly vulnerable spots, such as passages through walls or around corners. If you think that trouble might occur, choose a transformer with overload protection on the secondary (low-voltage) winding. It might cost a bit more, but it is a quality feature that could pay for itself by protecting the transformer from burning out. This feature cuts off current in the secondary winding if damage, such as a direct short circuit, occurs to the wires.

When you install a bell or chime, be sure to locate it where the sound will reach you anywhere in the house. The wiring is so simple that you do not have to pick a location to make wiring convenient.

For the chime itself, Figure 10-3 shows wiring to an eight-note, three-door unit. The circuit is not elaborate, but the chime can give a distinctive 4- or 8-note chime for the front door, one-note for the back, and a different single note for still a third door, if desired. Other combinations are available and, as we have said, can probably be installed with little difficulty by following the instructions included with the unit you purchase.

WIRING DIAGRAM

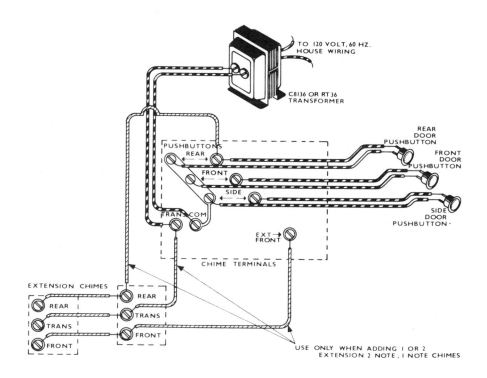

Fig. 10-3(a). Wiring diagram for a 4-8 note, three-door chime system. (Courtesy of Rittenhouse Division, Emerson Electric.)

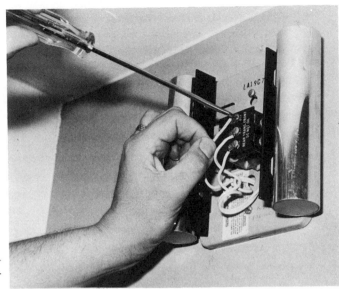

Fig. 10-3(b). At the chime unit wiring connections are made easily with a screwdriver. (Graf–Whalen photo.)

OTHER LOW-VOLTAGE GADGETS

Burglar/fire alarms, antenna jacks, intercoms, antenna systems, antenna rotators, outdoor decorative lighting (covered in Chapter 11)—almost an endless variety of systems and facilities can be added to your house through the use of low voltage. They are limited only by your pocketbook and imagination.

Many of these systems are available in kits that contain all necessary wiring, components, and power supplies. Many of the kits available will operate from batteries, but you could operate them just as well from the low-voltage chime transformer or from additional transformer units as required. You could probably install most of the available low-voltage systems by just following the manufacturer's instructions. But to indicate what might be involved, consider the circuit in Figure 10-4. This shows the basic circuit for the GE remote-control, low-voltage switching system.

Just as in the bell system, the transformer receives line voltage (115 V) and steps it down to 24 V in the relay control circuit. What is a relay, you may ask. Just by inspection, the figure can probably tell you. It is really an electrically-operated switch. When you press the ON button, 24 V flows through a coil in the relay and creates an electromagnetic pull that moves an armature to close a switch. We discussed electromagnetism in Chapter 1. The electromagnet works on an extension of the laws we discussed there. Electricity in a coiled conductor will produce a concentration of lines of force through the axis of the coil and cause the coil to become a temporary electromagnet.

As you can see in Figure 10-4, current in one coil moves the armature to close the switch while current in the other coil opens the switch. So what? Why can't you do this with a regular toggle switch? The reason is a matter of economics.

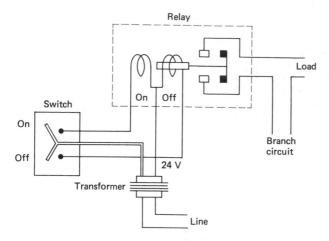

Fig. 10-4. Basic circuit for General Electric's remote-control, low voltage switching system. Relays handle the switching of the current. The relays are controlled by switches operating at low voltage that permit the use of wiring similar to that used for door chimes. Many different components are available. In planning and installing remote control systems, instructions supplied by the manufacturer of the equipment being used should be followed.

The circuit you switch on and off will no doubt be a 115-V circuit. Yet, you can control it with a 24-V signal over inexpensive light wiring, from 50 ft to 100 ft away (e.g., from the house to the pool area or outbuilding at the other end of your property) without having a special No. 14 or No. 12 cable running that entire length just for the switch. The 115-V circuit can have a parallel switch arrangement in the 115-V line, right at its operating location, but for a second or third control point, the low-voltage wiring will produce a real savings. Any number of switches controlling the same outlet can be easily and inexpensively added to the circuit. In addition, the use of master-switch assemblies allows control of many outlets from one or more locations. The arrangements they allow are literally endless. Check for the systems available in your area, read the instructions, and go to it.

YOUR OIL BURNER SYSTEM

Though not unusual because almost every house has one, the household heating system is a complicated and important element of your house. You should know as much about it as possible. Here again you can find all you need to know about installation from the manufacturer's instructions that come with the furnace. These instructions should be studied carefully before doing any installation or repair work.

The system wiring diagram will probably look pretty forbidding at first, but a little careful study should make it understandable. Turning the heat on and off by thermostat or other controls is simple enough. However, all the added interlocking safety devices that are designed to prevent accidental fire or explosion help to create an elaborate system.

A simplified oil burner circuit is shown in Figure 10-5. Basic parts are: the thermostat (TH); the transformer (TI), which steps down 115-V a-c line voltage to about 24 V to operate the thermostat; the relay (R), which is an electrically controlled switch; the motor (M), which operates the pump that brings oil from

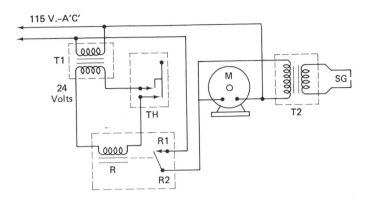

Fig. 10-5. This is a simplified diagram of an oil-burner electrical system as controlled by a thermostat marked by the symbol TH. Full details of the current cut-off and supply are given in the accompanying text.

the storage tank and forces it through an atomizer (vaporizing) nozzle in the oil burner; the ignition transformer (T2), which steps up the 115-V a-c line voltage to about 10,000 or 12,000 V; and the spark gap (SG), across which the high voltage arcs to ignite the atomized oil and produce an intense flame.

The thermostat is a temperature sensitive switch that can be set to close and open within narrow temperature limits (operation of a thermostat is covered more completely in Chapter 12). If the thermostat is set at 70° F, for example, and the home air temperature in the thermostat vicinity drops a few degrees below this value, the thermostat contacts close.

The thermostat is in series with the 24-V secondary winding of transformer (T1) and the coil winding of relay (R). When current from the 24-V transformer winding flows through this circuit, the relay coil becomes an electromagnet that pulls the armature so that two contacts of the relay complete a circuit in the same way as a switch does. In turn the relay contacts are in the 115-V line to the pump motor and ignition transformer (T2). Simultaneously, this turns the motor on to pump oil through the oil burner nozzle and causes a spark to arc across the gap of the ignition coil. Oil is atomized and sprayed into the furnace through the spark gap and is ignited by the spark plug. The spark is continuous but is really only needed for a few seconds to start the flame; beyond the first few seconds the oil flame continues on its own.

The flame heats or boils water in a boiler (in a hot water or steam system) or heats the air in a hot-air furnace. Some systems contain a second motor that operates a blower in a hot-air system or a water pump in a hot-water system to circulate the heat through the house through radiators of some sort. When the air in the vicinity of the thermostat heats up enough to affect the sensitive element, the thermostat contacts open and the pump or blower stops until the air cools off and the cycle repeats. The ignition transformer also shuts down.

There are many additions to this simplified system, which will, of course, de-

161

pend on what system you purchase. Look for the following types of controls on your diagram:

> Stack control switch: another heat sensitive switch that is placed in the chimney stack. In case the exhaust from the burner becomes too hot, this stack control will close down the whole system to prevent a fire hazard.

> Dual contact thermostats with heater elements that accelerate the thermostat action and cut the time the heating system operates. Without this, an inherent lag between the radiator system and thermostat location would allow the furnace to run excessively (remember, it takes time for the heat to circulate).

> A safety switch that will cut power to the furnace after a predetermined time (120 sec or so) if the cycle that was described previously starts but the oil does not ignite due to malfunction of transformer (T2) or a fouled spark gap.

> Hot water or steam control switches that will cut the circuit in case of excessive boiler temperatures.

Note that hot water systems sometimes have a different control mechanism entirely. The water-temperature sensing switch turns the burner on and off to maintain a preset water temperature, while the thermostat simply controls a secondary, motor-driven water pump to circulate hot water when the thermostat calls for heat. This has the additional advantage that the boiler water, which is kept hot, can be used to heat the hot-water supply. Gas-fired steam and hot-water systems are less complicated than oil systems because they do not need a pump motor or an ignition system. Instead of the oil pump motor, a magnetically-operated gas valve controls the gas flow. A small permanently-lighted pilot flame ignites the gas whenever it flows. In the case of failure of the pilot light, a safety valve that normally is held open by the heat of the pilot light closes down to prevent any gas from flowing. It cannot start again until the pilot light is manually relighted and the heat it produces reopens the safety valve to allow gas flow.

Another available variation is the timer thermostat. If you want to keep the house temperature lower during the night, buy a thermostat with a clock-timer as part of the mechanism. Most of these are designed with two thermostats in series so that you can set the two units at different temperatures. The day unit is always in the circuit, while the night unit is shorted out during the day. The timer opens this short circuit and puts the night thermostat into action at night and cuts it out during the day. If you decide to employ a timer thermostat, be sure that there are detailed instructions available when you make your purchase. If they do not seem complete enough, contact the manufacturer directly and do not proceed until you are satisfied that you have all the details you need.

EMERGENCY AND PORTABLE POWER

When a big storm blows up and utility lines go down, your home and business are suddenly robbed of the precious energy that powers refrigerators, freezers and heating systems, chases darkness with lighting, and keeps you in touch with the outside world. Memories of huge blackouts also call to mind the widespread looting of businesses made defenseless by the loss of electrical power. If that's one of your concerns, you'll be glad to know that you *can* connect a portable a-c generator to your home (or business) power wiring, to provide at least some of the essential energy needed until utility power is restored (Fig. 10-6). Generators typically are powered by a gasoline engine and convert the engine's mechanical energy to electricity.

Another benefit of some of these generators is portability. You can supply a 115-V 60-Hz ac to operate standard electrical equipment in your backyard or wherever you need it. Also, for camping, you can operate a TV, refrigerator, and other small units to which we have all become accustomed. In other words, portable or emergency generators can do a lot for you if you know how to use them.

Portable a-c generators are compact gas-driven power plants. The distinguishing features of portable generators are their compactness, low weight (ranging from about 60 to 130 lbs) and small engine ratings (about 3 to 7 hp). Output electrical power ratings range from 1500W for the smallest models to 3500W for the huskiest models. Some are compact enough to take with you in the trunk of a full-size car, to stow in a camper, or to keep handy in the garage or tool shed for those times when you need convenient power, or as a back-up generator to fur-

Fig. 10.6 A portable A-C generator. (Courtesy of McCulloch Corporation, Los Angeles, CA.) (Graf–Whalen photo.)

Fig 10-7. Standby Electric Transfer System connects power plant to home electrical system to provide emergency power. Transfer switch severs outside power company connection as it connects auxilliary power plan to house wiring. Installation is best left to a licensed electrician. (Courtesy of Sears, Roebuck and Company.) (Graf–Whalen photo.)

nish essential power if commercial power goes out in a storm. Larger models [such as the Sears Craftsman line (Fig. 10-7)] are available on wheeled carts, for ease in getting the generator into position at the connection point.

Power output in watts. This is the main factor to consider when buying a generator, and you should determine the need for the maximum power you will require well in advance of considering physical size. Table 3-1 will help. It lists the wattages consumed by various home and farm electrical devices. For emergency power you can probably get by with furnace, refrigerator and freezer, and a few lamps strategically placed about your house. If more than one will be powered at a time, add their wattages together. Next, add a safety factor of 20% to account for variations in load presented to the generator by different devices during start-up. The total wattage you add up will be the wattage needed. Buy the generator that has the nearest *higher* wattage rating. (That is, if you add up a need for a 1600-W output, don't try to get by with a 1500-W generator because the normal load exceeds the generator's capacity. Instead, buy a 2000-W generator. It will run smoother for longer and with fewer maintenance problems because it has reserve capacity beyond the needed power.)

The wide range of generators on the market can cover almost any need, but be sure the engine matches the power output you have decided that you need. 746W equals 1 hp; 1,000W (1 KW) is equivalent to 1.34 hp. However, to be practical, particularly in small sizes, efficiency losses of various sorts can cut this down a bit. It would be better to allow for 2 hp for each kilowatt of generator output. For example, a 10 hp engine would be sufficient for supplying 4 to 5 KW.

Noise level. In some cases, the engine that furnishes drive power to the generator will be running outdoors for prolonged periods. Choose a unit with an effective muffler system that reduces engine noise to the lowest practical level. If you will use it within 100 ft of your own house or that of your neighbors, get the quietest design you can find.

Fuel. A fuel capacity that is too low increases the number of times that the generator must be refueled. As a minimum, your generator should run for at least 90 min at rated power output on a single fueling.

Interference. You may wish to power radio, TV, or communications equipment from your generator. If so, be sure that the ignition system of the engine has the latest interference-suppression devices so that the generator does not become a source of electrical interference in the area of use.

Electrical outlets and extra features. To simplify connecting of the generator to the device it will power, built-in a-c outlets are provided. These may be the same as standard duplex receptacles found in your home's wall outlets in those generators designed to give a 115-V 60-cycle a-c output. Some generators provide more than one outlet and some offer both 115-V and 230-V output. But, remember that the total power delivered to all devices plugged into these outlets must not exceed the generator's output wattage rating. For short circuit protection, look for fuses or a circuit breaker. Also, if you have to charge batteries from time to time, you may want to consider a generator that provides 12-V d-c output, too. Some generators can provide up to 8 A of charging power for batteries in outdoor equipment, RVs, and remote equipment.

Starting. Some models have a fuel shut-off valve. This must be opened to allow fuel to reach the carburetor of the engine. (It is normally closed for safety, either when shutting off the generator engine or when the generator is in transit to prevent fuel spillage.) For easy starting, unplug any devices from the generator's outlets or turn their power switches off. (It is very difficult to start a generator when a load is connected.) Next, close the choke (if cold-starting) and operate the generator's starting mechanism (typically, a pull-rope recoil starter). Once the engine starts, allow it to warm up, then gradually ease the choke to the open position.

Running. Once the engine has come up to the proper speed, the generator will be providing full output to the outlets. You may now plug in or turn on the electrical devices to be powered by the generator. If current drain is substantial, you may notice a variation in the engine sound at start-up as the generator commences to supply power to the load. This should smooth out as current flow stabilizes. With motors on the line, your major problem will be the start-up surge. Some motors require three to four times the normal running current for start-up (see Table 10-1). It is good practice to turn on one motor at a time if you have several

TABLE 10-1. Power Requirements of Electric Motors

Motor hp Rating	Approximate Running Watts	Split-Phase Motors	Capacitor-Start Motors
1/6	275	2050	850
1/4	400	2400	1050
1/3	450	2700	1350
1/2	600	3600	1800
3/4	850	. . . .	2600
1	1100	. . . .	3300

on your circuit. Let the first come to speed before cutting in a second motor; start big motors first, then smaller ones. Depending on your system capacity, you may even have to turn off all lights or smaller items before switching in any of the motors.

Connecting your generator to your home wiring system. Connection of the emergency generator to your home system is, of course, the heart of the problem. When utility company power goes out, do you simply plug your generator into the fuse box and run the circuits you want? What happens when line power suddenly comes back? It could blow out your generator because it is not in phase.

To avoid this, you need an all-purpose transfer connection that cuts in your generator and cuts out the commercial power line until you are ready to reconnect. This connection is called a *transfer switch* and is installed as shown in Figure 10-8a between the utility company meter and your service entrance. If your meter is outside, the transfer switch will be at a convenient point for generator connection, but it must be installed in a weatherproof box.

As you can see from Figure 10-8b, with the switch in normal position, utility company power feeds right through to the service entrance panel and the generator connection is out of the circuit. In emergency position, the power line is cut off completely and the contacts to which you connect your generator are switched into the circuit. Remember, though, before you throw the transfer switch the emergency generator motor should be up to speed and running smoothly. Just as important, all unnecessary branch circuits should be turned off or disconnected and all unnecessary lights and appliances on the emergency circuit should be off or out of the circuit to avoid a sudden load on the generator. It would be well to have a note of some sort on the service entrance describing just what circuits to turn off and in what sequence equipment should be turned on after connection of the generator is made.

You may wonder what will happen when power is restored. The answer is: *Nothing.* Line power will be disconnected until you reconnect it manually. You

Fig. 10-8. (a) Transfer switch between service entrance and utility company meter. (b) Transfer switch (left, above) takes power from system meter input, heavy cables atop, or from generator, light cables at bottom of switch. Depending on switch position, one or the other supply feeds into the main distribution box. (Graf-Whalen photos.)

will have to keep track of street lights, neighbors, and radio reports to tell you when your commercial power is back. Turn off all equipment (disconnecting all branch circuit fuses or turning circuit breakers off at the service entrance would be effective), reverse the transfer switch, then add circuits to the line a circuit at a time. Not that your particular house is not likely to cause your utility trouble, but if everyone in town turns all their circuits back on at once, you are liable to cause another outage.

RULES FOR SAFE OPERATION OF YOUR PORTABLE A-C GENERATOR

Never operate the generator in an enclosed space. Provide adequate ventilation. Gasoline engines give off deadly carbon monoxide, which can be lethal if inhaled.

Never fill the gas tank while the engine is running or in the dark. Gasoline spillage on a hot engine can cause fire or an explosion.

Alternators produce lethal voltages and should be treated with respect.

Comply with laws governing the storage of gasoline.

Maintain power cords in good condition. Bare or frayed wires can cause dangerous electrical shock.

Check engine oil level each time you fill the gas tank.

Never attempt to change engine speed without proper knowledge and equipment. Incorrect engine rpm is not only dangerous but can damage the alternator or the equipment it powers.

Study the contents of the manufacturer's manual carefully before operating your generator.

11

OUTDOOR WIRING

Porch, Yard, And Pool Lighting—Low–Voltage Systems

As the sun slips below the horizon and darkness enfolds your house and yard each evening, you can bring instant beauty to your home by turning on a carefully planned, colorful outdoor lighting system. Safe, economical to operate, and easier to install than you might think, today's weatherproof outdoor lighting systems let you landscape with light—spotlights can be used to splash light of different hues about your house and yard, bringing out other-wordly qualities in trees and plantings. Colorful lights can illuminate pools or garden waterfalls and fountains sprayed by hidden electric pumps. Area lights also can shed a cheery, festive glow around patios and along walkways, making after-dark entertaining a pleasure. Moreover, a well-lit home is a more secure home. Burglars steer clear of homes shielded by a protective, surrounding aura.

Lamps can be mounted in plastic housing of two kinds: upright spots, with push-in ground stakes to hold the light erect, and mushroom fixtures in which the light shines up against a translucent or reflective dome that sheds even area illumination on pathways or plants. Wall and post mounting are also available for added flexibility in setting up a lighting scheme.

The average household 115-V circuit is fused at 15 A. This means that its capacity is 1,725 W. To see if an existing branch circuit will handle your lighting and other outdoor needs, add up the wattage of all lights, motors, water fountain pumps, and so on that will be used *at one time* and powered by the same line. If this totals more than 1,725 W, a separate circuit should be run from the house main panel.

Underground 115-V wiring with wire of types UF or USE can now be used without enclosing in conduit or lead sheath. Where the cable emerges from underground, it must be protected by thin-walled Greenfield (flexible) conduit secured to a post. In all cases, *this outlet must be grounded* and protected by a GFCI. Weatherproof outlets can be installed on fences, posts, trees, or buildings. You may also wish to have an inside switch for your system as dusk approaches.

PLANNING YOUR LIGHTING SCHEME

Consider just what lighting effect you are trying to achieve. If your idea is to illuminate trees or plantings, spotlights are the logical choice. Here, the two major lighting methods, *uplighting* and *downlighting*, can make your property's trees a dazzling spectacle.

For uplighting, one or more spotlights (spots) are staked in the ground below a tree and directed up into its leafy branches. This produces an everchanging pattern of illumination and silhouetting as breezes stir the leaves, providing an enjoyable sensory experience for viewing from your patio.

Downlighting of trees requires placement of spots in upper branches so that the light shines down through the leaves, colorfully illuminating the ground below. Shifting patterns of color and shadows are thus projected onto the ground surface, providing an inviting, soft background for leisurely conversation or strolling.

Plantings or outside statuary can also be illuminated by colorful spots, carefully placed so that the light source is not apparent to the observer. If the color of your blooms is the thing you're trying to preserve and display, choose the color of the light cast upon your plantings with particular care. A red light will dramatically enhance the color of your roses—but the greens around it will show up a muddy brown! A green light is best for foliage, but may distort the hues of bright blossoms that are orange or red. If your objective is to enjoy the natural beauty of your garden by night as you do by day, choose a white- or bluish-white light source to illuminate your plantings and splash colors elsewhere.

A pool or patio can be ringed with light by using "mushroom" lamps providing area downlighting (see Fig. 11-1). The soft, diffused light shed by these lamps can be augmented by spots illuminating nearby features of interest. Try to avoid lamp placement that subjects people within an area to the direct harsh glare of a spot. This creates a feeling of being "on-stage" and also interferes with the comfortable conversational grouping of guests. Try to locate your lighting so that reflective light falls on the area of activity. It is more flattering, congenial, and decidedly more comfortable!

Falling water and colorful lighting are natural partners in creating a striking exterior decor. A fountain or waterfall operated by submersible pumps and illuminated by spots or special underwater lamps becomes a scene of rippling, brilliant, dynamic beauty. Avoid glare; your placement of light should carefully avoid lighting angles that bounce annoying reflections back at the viewer.

Lighting of a house exterior must be planned with extra care to preserve the special qualities that make your home a cherished dwelling-place. Low-key color

Fig. 11-1. Outdoor lamps with mushroom light shields are easily installed and provide non-glare lighting for paths, walks and pool areas. (Graf–Whalen photo.)

can be added by area or walkway downlighting with mushroom lamps. Spots placed in soffits under the eaves or on nearby trees can shed soft downlight on lawn and plantings immediately in front of the house, creating a subdued reflection of color onto the house itself. Avoid spots on the lawn shining directly on the house; the harshness of this type of lighting suggests the appearance of a model home in a new development. Try the more subtle approach of indirect illumination. It reflects better upon you as a proud homeowner.

PLACING THE LAMPS

Spots may be affixed to short stakes if ground-level placement is desired or to long stakes if you wish to locate a light above surrounding plantings. Alternatively, you can secure the spot mounting bracket to a wall, shingling, fence, or tree using appropriate fasteners. Angle the mounted spot to create the desired lighting effect, then lock the light at this angle (many spots have a wing-nut clamp that allows fast, positive adjustment using finger pressure only). Mushroom lamps have attached stakes for insertion into soft earth adjoining walkways. Be sure to leave at least 18 in. clearance between the lamp's dome and path edge.

WIRING YOUR SYSTEM

First let's consider 115-V circuits for outdoor use. The simplest outdoor wiring consists of a weatherproof box with a dual receptacle for extension-cord plugs mounted on the outer surface of your house. Cable or conduit to the outlet can run from an interior junction box, often in the basement area, and right through the wall to outside.

It goes without saying that the circuit you use will be overcurrent-protected. However, for outdoor work, grounding is also critically important. You could easily contact an outlet box or switch while standing on damp ground or on wet pavement near a pool. As we describe in detail in Chapter 2, this is a potentially hazardous situation because the resistance between your body and ground is so low. For several years now, the NEC has made the installation of a GFCI a requirement for all outdoor circuits other than low-voltage circuits.

As we have mentioned before, GFCIs may be built into the service entrance panel as GFCI circuit breakers (Chapter 4) or built into one of the first boxes of a circuit as a feed through receptacle to protect all added boxes on the circuit. Either of these methods would provide adequate protection for an outdoor line. However, if neither type circuit is available, the simplest way to protect yourself is to plug a portable GFCI dual receptacle into your outdoor weatherproof box as shown in Figure 11-2. These plug into the outlet, and extension cords are plugged into them. They do afford protection but are only intended for occasional use, such as running tools or other devices outside. Because they are not a permanent installation, they will not meet code requirements for permanent outdoor circuit protection.

Even with the GFCI protection, any conduit used should be grounded to the house ground bus or you must use cable with ground wire and ground your outdoor receptacle or GFCI to the box with a grounding jumper. The weatherproof box must be equipped with a cover, generally spring-loaded to stay closed when not in use, and must include a gasket to seal the cover to the box and prevent moisture entry. Most weatherproof boxes also are constructed with the top surface slanted out and downward to act as a rain shield for added protection.

This type of installation is very handy for supplying power to hedge trimmers, lawn mowers, and so on by means of a grounding-type extension cord. It is,

Fig. 11-2. With the extension cord of the GFCI box plugged into a receptacle in the garden or garage, two separate extension cord circuits can be afforded GFCI protection. (Courtesy of Pass & Seymour, Inc.)

however, limited in scope, as it is fixed at one point of your house perimeter. A more extensive system would probably make use of buried cables to supply outdoor lamp posts, pool lighting, cabanas, or the like at a greater distance from the house. These systems *must* start from GFCI-protected house circuits.

UNDERGROUND WIRING

Weatherproof outlets and switches and wire made for underground runs are becoming increasingly available. However, when planning to use underground wiring, check your local code carefully for the type of wire they require. Sometimes *U-F* wire (underground-feeder) is used unless specifically prohibited by the local code. *U-F* is impervious to moisture, rot, and dampness (BX, which would rust, is strictly prohibited). The cable must connect to a GFCI and breaker-protected circuit inside the house. It can extend outside through foundations or walls as shown in Figure 11-3. Your local electrical supplier can probably advise what type of fuse box to use inside.

Underground cable should be buried at least 24 in. deep to prevent damage from frost or from accidental contact with deep-cutting garden tools. If a lawn area has to be disturbed, you can generally cut the grass away. Use a spade and set the grass sod carefully to one side of the trench before digging deeper into the soil. The soil removed to a depth of 18 in. or so can be set to the opposite side of the trench. Note that the last 6 in. of depth can most easily be achieved simply by digging the spade into the bottom of your trench and wiggling it side to side to form a slit. This will normally be wide enough for the cable to be installed. Do not pull the cable taut in a trench. In fact it is good to snake it side to side to allow for expansion and contraction of the ground. After the cable is installed, fill in the trench and replace the sod. It might also be a good idea to place a redwood board in the trench above the wire to protect it from a sharp tool such as a garden spade that might cut the wire accidently.

The size of the wire you use will depend on the current you want the circuit to handle. Most outlet receptacles and yard-light circuits would be fused for 20-A service and use No. 12 two-conductor with ground cable. But if you are running it some distance from the house, remember to increase the size, probably to No. 10, to prevent voltage-drop problems. Also, for large jobs such as a central air conditioner mounted outside the house, use even larger wire, depending on the rating of the unit.

If you would rather not run the wire underground, you can run it overhead from building to pole, and so on, but you must maintain a minimum above-ground clearance of 12 ft. Until recent years this was the normal practice, though it has disadvantages. Even if you are not bothered by the appearance, consider the other problems; ice, wind, and falling trees or branches can easily destroy your work.

Where your cable leaves the ground, it is particularly vulnerable and must be protected accordingly. Usually the best protection is a conduit around the wire

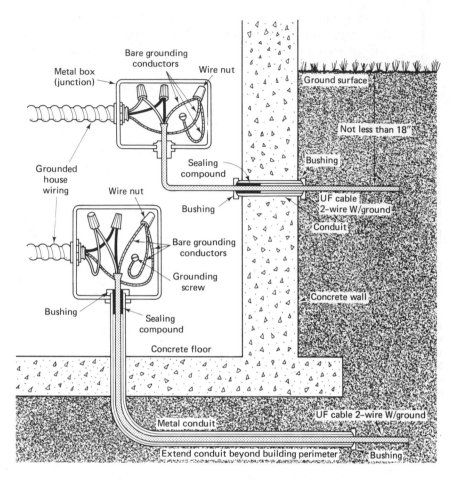

Fig. 11-3. Conduit must be used to protect UF cable entering a building through a wall or a floor. Courtesy of *Popular Science;* Copyright 1974 by Times Mirror Magazines, Inc.

wherever it is exposed and accessible. The conduit should extend at least halfway down into the trench and continue upward into any exposed electrical fixture (see Fig. 11-4). A plastic bushing should also be placed around the cable where it enters the conduit underground. This will prevent any possible sharp edges of the conduit from cutting into the cable insulation.

Water is one of your main hazards outdoors. Keep the conduit as vertical as possible where it leaves the ground to prevent trapping water at the cable. For sockets, switches, and lights, be sure you use only approved outdoor cables. Built-in gaskets and seals on outdoor components prevent entry of water to the vulnerable wire connections inside.

Connection rules are quite similar for outdoor and indoor work—connect black wire to black, white to white, ground to ground; connect black or red wire

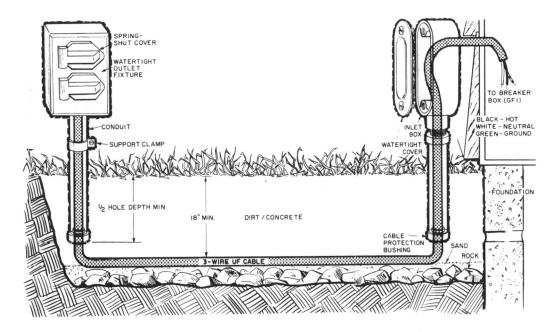

Fig. 11-4. Simplified run shows the mechanics of an underground cable installation. Cable must be protected over its entire length against possible damage and moisture. For the underground run, a minimum of 18 inches of soil must cover the UF. (Twelve-inch depth is okay if a concrete covering is used.) If burying in rocky soil, place cable on layer of sand for protection. Bushings protect cable from possible sharp edges of conduit. Above ground, conduit must cover the cable wherever it is exposed. Approved watertight fixtures must be used throughout. Courtesy of *Popular Science;* Copyright 1974 by Times Mirror Magazines, Inc.

only to brass-covered screw terminals and white to silver-covered terminals; connect switches only in the hot line.

Porch or door lights are usually too far from the street to illuminate small obstacles. For this reason you may want to install a post lantern on a driveway or walk some distance from the house. The method of installation of the post itself will depend on whether it is made of a wood or metal, but wiring is pretty straightforward. Lead your wire through the underground trench into the post underground and out through the top of the post. Connect the cable wires, including ground wire, to the wires of the fixture.

Figure 11-5 shows a weatherproof junction box used to supply several weatherproof dual outlets on a single circuit. Notice the sections of conduit that support both the junction box and outlet boxes to keep them off of the ground. Some codes will require that these be attached to posts or to the side of a house structure. Wiring within the boxes is identical to a similar circuit inside the house.

LOW-VOLTAGE OUTDOOR LIGHTING

Decorative and useful as it is, outdoor lighting and other effects need not be costly with low-voltage power available. A six-light set of walkway lights may use only a few cents worth of electricity in an hour, no more than a standard 100-W

Fig. 11-5. A weatherproof junction box. (Graf–Whalen photo.)

bulb, and still give broad coverage. Operating on a mere 12 V, this system can provide a great deal of illumination and still be safe. Low-voltage outdoor lighting systems may use automotive-type bulbs in tough reflectorized enclosures, or sealed beam lamps similar to those used in car headlights. Also, for pool or garden fountain lighting, submersible 12-V lights are available in many colors and styles.

Ease of installation is one of the most important advantages of this equipment. You can safely run wires almost anywhere you want; above ground, along fences, into bushes and trees, even through water. You can also run the wire as is, without digging trenches, uprooting lawns, or laying conduit. Also, the greater safety of these lights around a pool area makes them an ideal choice.

Low-voltage lighting is generally available in kits containing the required transformer, wiring, and all components ready to be mounted and strung. You can purchase 2-, 6-, or 12-light sets or separate components (see Fig. 11-6). Your only difficulty might be matching the system to the transformer. Be sure that you check the wattage rating of the transformer and do not exceed the recommended rating with the lights you use. Your supplier can surely help you in this calculation.

Most low-voltage lighting systems include an additional supply line of 100 ft, plus 25-ft cables on each lamp. This combination affords considerable flexibility in laying out the lighting system for your home.

The weatherproof transformer is usually installed in a protected location, such as a corner or behind stone-work, where it is not likely to be bumped with a power mower. The power cable attached to the transformer should be wired into a nearby, weatherproof outlet. Do not use an extension cord! Their use can cause lethal consequences outdoors and they are in violation of almost every electrical

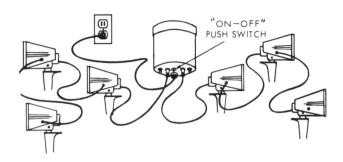

"ON–OFF" PUSH SWITCH

Fig. 11-6(a). A low voltage system such as that shown comprises a transformer that can plug into a 115 V outlet and supply safe 12 V power to a number of lamps. (Courtesy of Cable Electric Products, Inc.)

CAUTION: Do not connect more than three lampholders to each transformer outlet, as shown above.

code when permanently used in an outdoor circuit. Your best choice is to mount the transformer inside a weatherproof outlet box. Make 115-V primary connections directly to the 115-V house-system wiring and extend only the 12-V secondary wires outdoors. For adequate protection against moisture and inquisitive animals, mount the transformer outlet several feet above ground level.

Next, uncoil the 12-V secondary supply cable, laying it flat on the ground to determine how far your system's lights can be located from the transformer site. Some systems have a single 2-wire supply line, while others provide two 2-wire lines with separate connectors at the transformer. The latter type places the transformer at the center of the system, offering a somewhat easier layout of a lighting system to cover a front yard or to encircle a pool or patio.

Once the system wire is uncoiled and laid out, unwrap and lay out the 2-wire cords of each lamp to be used in your system. The 12-V cable leading from the transformer to the lights *can* be left on the ground surface, but it is safer to bury it. This will prevent accidental trip-ups and inevitable breaks by mowers and garden tools.

A good technique for burying the cable is the use of an edging tool to slice a narrow trench in the earth at a depth of 3 to 4 in. Once the earth is incised, use the tool to lift out the divot. In a short time, you will have a neat, inch-wide trench that is easily closed up later. Lay the wire into the trench, feeding toward the first light's hook-up position. When you are opposite the right spot, connect the lamp wire to the feeder cable. Lay the wire into the branch trench leading to the lamp. If you have more wire than is needed, dig a large hole about midway along the branch trench and drop in the excess. Feed out the other side, continuing through the branch trench. Position the light and temporarily ground-stake or secure it to tree or wall. Continue on to the next positions until all lamps are approximately placed in accordance with your light plan.

Most lamps feature clip-on connectors, having two insulation-piercing points and a screw-on fitting. The connector is simply placed over the 2-wire 12-V line from the system transformer, then screwed down until the two piercing points bite into the supply wires. This simple connection method makes for a quick hookup when you have found the right location for a light or rapid relocation if you are not satisfied with the effect gained on the first try. There is no need to patch the tiny holes left in the cable if a connection is moved.

12

REPAIR

Test And Troubleshooting—Fluorescent Lamps, Motors,
And Appliance Repair

Let's discuss maintaining and repairing your electric plant and all the parts attached to it. If you have installed it, you have a fair idea of how your system operates. With a few explanations, you should be able to extend that understanding to the plugs, lamps, motors, or appliances that your system services.

The first rule of repair is, of course, to turn off the power to the problem device or area. If a lamp switch fails to work, pull out the lamp cord and replace the switch at leisure. Should a furnace motor fail to turn, disconnect the branch circuit at the service entrance, then check the motor and wiring for a fault. If any appliance, switchplate, lamp, or fixture gives you a shock, immediately disconnect the power and look for the cause. While testing and troubleshooting, you may have to reconnect power occasionally. Be sure to do this carefully, warning others what you are doing and proceeding with caution.

Test Methods

Small, inexpensive types of testers will be quite helpful in solving your test-equipment needs. The tester shown in Figure 12-1 is a neon lamp in a plastic housing (about as big as a person's finger) and is equipped with two metal probes on the ends of short (about 6-in.) insulated electrical leads. One of the leads is red and the other is black. The red probe is for connection to positive or high voltages and the black for negative or neutral, depending on whether d-c or a-c currents are being tested.

In testing the receptacles of an outlet for the presence of voltage, hold only the plastic or rubber covering of the lead and bring the metal probes into contact with

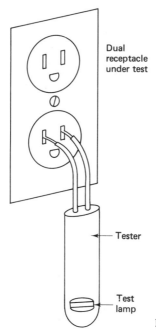

Dual
receptacle
under test

Tester

Test
lamp

Fig. 12-1. Voltage tester.

the points to be tested. First, insert both probes into the parallel blade holes of the receptacle. If the neon lamp lights, there is power in the circuit.

Next, check to see if the outlet is grounded. For a three-hole receptacle, the ground connection hole as well as the cover plate should be grounded. Insert the probes into the two parallel holes of the receptacle; the lamp should light to show that there is power. Then insert the black probe in the ground connection hole and the red probe into the other two holes in turn. The lamp should light with the probe in one of the holes; this is the hot connection. Leave the red probe inserted and remove the black probe to touch the cover (or screwhead). The lamp should light to prove that the outlet box is grounded. Several other tests of this same sort, called go/no-go tests, can also be performed with this handy lamp.

Figure 12-2 shows a more elaborate tester (a receptacle analyzer) that fits a single-phase, three-wire, 120-V outlet. This type of tester has three blades to fit the three-wire outlet and contains several neon lights that indicate instantly whether the circuit is open or live, whether the ground, hot, or neutral wire is open, or whether the hot, ground, or neutral lines are correctly installed. These analyzers cost a bit more but do so much with a simple insertion into the outlet that they are very popular. Read the instructions with these analyzers carefully and you will see how helpful they can be to you. Remember, though, as handy as these testers are, they are limited in use to circuits that are energized.

The volt-ohmmeter (VOM for short) is shown in Figure 12-3. It is a godsend for the more elaborate testing you may have to perform. Most VOMs have a single-needle pointer that sweeps across a dial with many scales. Switches on the

5200

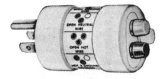

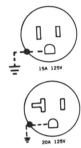

QUICK, EASY OPERATION

Neon lights indicate fault conditions, see chart.
Example: By plugging into single phase, 125V.,
2-pole, 3-wire outlets, left, the circuit condition
will be indicated by combination of lighted or
unlighted lamps.

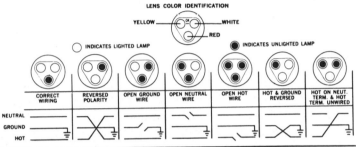

MAY BE USED TO TEST OTHER OUTLETS

(with polarized adapters only)

The 5200 Tester may be used to check many other
2 and 3-wire outlet types wired for single phase,
125V. service by use of proper adapters.

Fig. 12-2. Outlet circuit tester and polarity indicator. Courtesy of Hubbell
Division, Harvey Hubbell Incorporated, Bridgeport, CT.

housing allow selection of voltage or resistance as the quantity to be measured by
the pointer position. Other selector switches allow the scale readings to be varied.
Depending on the actual settings of these switches, the operator must take care to
read and interpret the proper scale on the dial (see Fig. 12-4).

To measure resistance, the resistance selector switch may be set at RX 1,
RX 10, RX 100, and so on. When the pointer is positioned, for example, at 2 on
the resistance scale, it can be interpreted as $2\,\Omega$, $20\,\Omega$, or $200\,\Omega$, depending on
the resistance-scale switch position. (Each position makes the scale readings ten
times higher than the previous switch position.) Just like the lamp tester described
previously, VOMs are equipped with metal probes on the ends of two insulated
leads, one red and one black. Often the leads are several feet long for user
convenience, and meters may be equipped with interchangeable leads with dif-

181

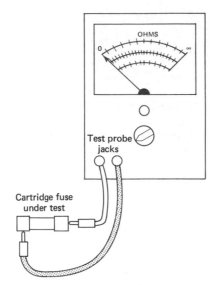

Fig. 12-3 With the bare ends of the test probes touched to the metal end caps of the cartridge fuse and the meter selector switch in the resistance position, the meter will read zero ohms if the fuse is good and infinite ohms, showing open circuit, if the fuse is blown.

ferent end probes that can all plug into the meter at will. Some probes are short, exposed metal needles, while others may be spring-loaded metal clips that can snap onto a conductor for firm contact. The red and black color-coding of the leads is important only for testing polarized circuits. (A polarized circuit has one positive side and one negative side, e.g., a d-c circuit.) In this type of circuit, the red probe is connected to the positive side and black to the negative side. The probes must not be connected in reverse or else the meter pointer will sweep backward and may actually be damaged if the voltage being measured is high. In an a-c circuit, either probe may contact either side of the circuit because ac is nonpolar.

For other tests such as resistance tests (which we will consider first), the probe connection is not important. They may even be interchanged after taking a reading and cause no change to the value. Before the actual test work, we must discuss another feature of the VOM; *zero adjustment*. Zero adjustment is normally performed with a ZERO ADJUST knob on the meter front housing and is done prior to resistance tests. First, the two meter probes are held together in firm metal-to-metal contact. This should give a 0-Ω continuity reading on the dial scale. If the pointer is not exactly at 0 on the dial, the ZERO ADJUST knob is then adjusted until the pointer lies exactly on the zero mark. All other pointer positions will then be accurate. You can learn a great deal from simple and safe resistance and continuity tests (see Fig. 12-3) with power off or with an inoperative appliance disconnected from the outlet. The battery contained in the VOM provides a low voltage necessary for these tests and you do not have to use 115 V while making measurements.

Remember that for current to flow, there must be a complete continuous circuit. When this exists the circuit is closed, current flows, and there is continuity. Any opening at a switch, a poor connection, or a break in a wire or component will open the circuit and prevent current flow. To perform a continuity test, first set the meter selector switches to resistance and RX 1. Then connect the meter probes to both ends of an electrical conductor (red or black probes can be at either end since this test involves no polarity). If the conductor has no break in it, the probes will complete a circuit through the meter and the meter will indicate about 0 Ω or continuity. The continuity test confirms that there are no breaks in the circuit conductors. This is good to know, but let's carry it a step further and discuss how to check the total electrical resistance of the circuit under test. For example, you have easy access to the power leads of a toaster just before they disappear into the unit. A continuity test will quickly show only if the circuit is continuous. But, we know that a toaster coil should have a resistance element of specified resistance so that it will impede current and heat up (see Table 12-1 for typical values). Suppose we perform a resistance test

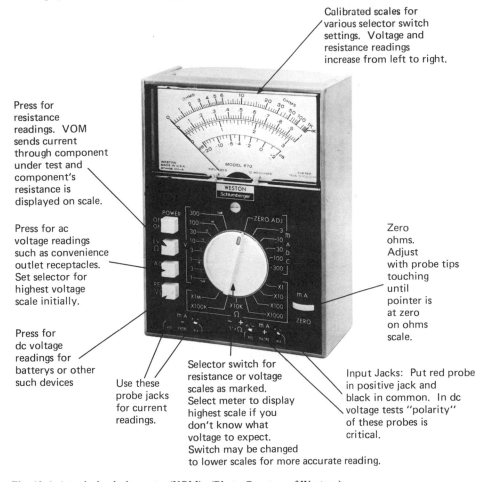

Calibrated scales for various selector switch settings. Voltage and resistance readings increase from left to right.

Press for resistance readings. VOM sends current through component under test and component's resistance is displayed on scale.

Press for ac voltage readings such as convenience outlet receptacles. Set selector for highest voltage scale initially.

Press for dc voltage readings for batterys or other such devices

Use these probe jacks for current readings.

Selector switch for resistance or voltage scales as marked. Select meter to display highest scale if you don't know what voltage to expect. Switch may be changed to lower scales for more accurate reading.

Zero ohms. Adjust with probe tips touching until pointer is at zero on ohms scale.

Input Jacks: Put red probe in positive jack and black in common. In dc voltage tests "polarity" of these probes is critical.

Fig. 12-4. A typical voltohmmeter (VOM). (Photo Courtesy of Weston.)

TABLE 12-1. Resistance Values of Common Appliances and Motors

Appliance	Ohms
Radio	5
Clock	600 to 1200
Mixer	7
Toaster (two-slice)	12 to 18
Waffle iron	8 to 10
Deep fat fryer	10
Rotisserie	9
Coffee maker	24
Laundry iron	14
Heating pad	300
Electric blanket (double)	100
Vacuum cleaner	1 to 2
10-inch fan	22
Electric knife	40
Electric can opener	8
Shaver	140
Hair dryer	25
1/4-inch drill	10 to 20
1/2-inch drill	4 to 9
1/3-hp drill-press motor	1-1/2
7-1/2-watt lamp	400
60-watt lamp	5
Small soldering iron	700
Large soldering iron	85
Gun-type soldering iron	7
Belt sander	3 to 9
Portable circular saw, 7-inch	4
1/20-hp split-phase motor	7
1/4-hp split-phase motor	1 to 4
1/3-hp split-phase motor	0.8
1/3-hp capacitor motor	1.5
1/2-hp capacitor motor	0.6 to 1.2
3/4-hp capacitor saw motor	1

and find that the resistance between the two leads is near zero; this would mean that there is a short circuit within the toaster, which is worse than a break in the coil. Besides not working, the faulty toaster would allow very high current to flow, which could possibly cause a fire.

After these go/no-go tests, you might also want to check that the actual resistance of a component is at a specific value, plus or minus a small tolerance. For example, knowing the normal resistance value of appliances and motors (see Table 12-1 for typical values) will help you to decide whether a unit is defective because many wire turns in a winding have shorted together and reduced the resistance to far below that of the normal operating value. It is interesting and instructive to learn to use the VOM and check several small appliances, lamps, switches, and so on (disconnected from the circuit, of course) to familiarize yourself with normal values. Then, when you actually work on a unit you suspect of trouble, the work will be second-nature to you.

When the resistance being measured is high, the meter can be set for the highest ohm scale. The meter can even measure the resistance through your body as you hold the test probes in your hands. Where the measured resistance is low, the meter can be set on successively lower resistance scales until the low resistance gives a large pointer deflection and the resistance can be read accurately. Low resistance tests can be made on the RX 1 scale. Generally speaking, for most accurate reading, choose the scale range that has a mid-scale reading at about the resistance you expect to check. Touching both probes to two ends of a motor winding, for example, should deflect the meter needle to read the resistance of the winding. Just be sure to read the resistance on the scale marking that corresponds to the selected scale on the switch.

In troubleshooting, it is a good idea to start at the line cord, the point where power enters the appliance. The VOM shows a break on portable appliance cords by indicating an open circuit. If there is no break, the meter should read continuity (the needle will sweep upscale toward 0 Ω) when you touch one probe to one blade of the plug and the other to the other end of that wire where it enters the appliance. Similarly, check the other blade of the line cord. If you get low-resistance readings in both cases, the cord is probably good. Also, check the ground connection, if provided.

If you still suspect that the cord is damaged, attached fasteners to each test probe (or use the clip probes already equipped) and clip the probes to the wires as before. Then bend and twist the wire to determine whether there is an intermittent break in the conductor. If a break exists, the needle will drop momentarily to the left and show a high resistance reading. When a cord must be replaced, get the same type of cord with a wire at least as large as the original.

The next most-likely culprit in malfunctioning appliances is the switch. Because it opens and closes the circuit and operates many times in normal service, it can easily fail. In the closed position, the switch should show 0 Ω (continuity) between the two terminals; when it is open, the resistance should be infinite.

Another area of possible failure is the fusible link, incorporated in some appliances to protect against overload. These devices melt to break the circuit and protect the lines if an overcurrent situation occurs (in much the same way that a fuse operates). Quite often, it is almost impossible to check this link visually and the continuity tester is the only way to detect this cause of failure.

The load is that part of the appliance that does the work. It may be a motor, a solenoid, a heating element, or any combination of these devices. Unlike the switch or cord, the load has a certain amount of resistance. Your VOM can be an invaluable aid to locating failures in the load. If something happens inside the load to cause the resistance to drop to 0 Ω (a short circuit), the appliance will generally overheat and smoke due to burned insulation unless it blows a fuse or trips a circuit breaker first. Burning insulation is generally discovered without a meter because the odor is quite strong. However, breaks in the load circuits (which are more likely to occur) are just like breaks in the extension cord and can be found by means of the VOM.

Checking an energized circuit for its voltage can be done by setting the proper selector switch on the meter. You have to be careful to avoid a shock. Test the circuit wires only with the tips of the probes and keep all other parts of your body clear of contact with the appliance you are testing. Two types of voltage measurements can be made, either a c or d c. For d-c tests, the red probe must be connected to the high (positive) side of the circuit and the black lead to the low (negative) side. For a-c tests, it does not matter where the probes are connected. For a voltage test, set the VOM switch for a voltage scale that is slightly above the value you expect to read; using the a-c scales for a-c house line voltages and d-c scales for battery-operated units. Clip test leads to the circuit or the portion of the circuit being tested, then turn the current on. Be sure to read the proper scale. If the correct voltage is read across the part but it still malfunctions, then the trouble is probably in the part, not the supply. No voltage reading indicates that the trouble is somewhere in the supply network. If you plug in a disassembled appliance, remember that the exposed leads, heating elements, and other parts are live wires. *Treat them with respect—always.*

There is one more test you can make with your VOM. Even when an appliance works, current sometimes leaks to the ground frame of the unit. This sometimes happens in motorized appliances when insulation ages and allows current to leak to the grounded case. This can create excessive current, causing excessive heat in the working element and failure. In an ungrounded appliance, this will create a shock hazard. Make it a habit to test the housing of the appliance for proper grounding after testing for continuity. To make the test, set the range selector switch to the highest resistance scale RX 10K. Then test from the metal housing of the device to each terminal where the cord attaches. A no reading (infinite resistance) should be noted; the needle should stay at the extreme left end of the scale. If you notice a slight needle movement on the RX 10K scale, change to a more sensitive scale and see if you get a reading. Any measureable resistance indicates that the insulation has broken down somewhere in the unit. If the defect cannot be located and repaired with insulating tape or the like, the motor or other major load device will probably have to be replaced.

For a grounded appliance with a three-blade plug, add one more step to this ground test. Touch one test probe to the ground blade of the plug and the other to the appliance housing. You should read $0\,\Omega$, which confirms that the case ground connection is good. If current leakage should occur, the ground connection will protect a user from shock.

TROUBLESHOOTING AND REPAIR

Fluorescent Lamps

Fluorescent lamps, those marvelous energy savers, are often subject to mysterious ailments. Flickers, hard starting, quick start and then going out—all of these problems can occur.

If the tube appears dead and will not even flicker, check the tube. Turn it gently in the socket to be sure contact is good. Inspect the ends near the electrodes. A lamp nearing the end of its useful life will have blackened tube ends extending for about 2 in. from one or both electrodes. This is caused by erosion of the active electrode material. Each time the tube is switched on, some material erodes and deposits on the inside of the tube. If the tube looks heavily blackened and the contact appears alright, try another tube from a similar fixture and check its operation; chances are your tube has reached the end of its useful life.

Now, what do you do if the replacement lamp doesn't work either? Replace the starter with one of identical rating and try again. It is always a good idea to keep a spare starter on hand or, if you must, borrow one temporarily from a working fixture that is identical to the defective unit. The starter is removed by first pressing it in slightly, then turning it counterclockwise half a turn until it pulls out easily. If the replacement starter does not do the trick, turn off the circuit, remove the fixture, and check the wiring connections. Check the ballast for a short circuit between any sets of contacts (you will need the fixture schematic and your VOM). Check the switch for proper operation.

Other troubles can be traced in the same general way as follows:

1. If the lamp blinks on and off, it may be near the end of its useful life. Check the lamp. Try replacing the lamp. If that does not help, check for a loose connection in the fixture—with the circuit dead, of course.

2. Swirling or spiralling light along the tube in a new lamp is nothing to worry about. Switch the light on and off rapidly a few times and this trouble should disappear.

3. If the end filaments continue to glow after the lamp lights, there could be a shorted capacitor in the starter or starter contacts welded together. In any case, replace the starter.

4. If the lamp is hard to start and the lamp checks out good, the starter may be defective or near the end of its life. Replace the starter and recheck. If starter replacement does not work, replace the ballast.

5. A steady or intermittent hum in the fixture may be caused by a ballast hum, which is normal; the hum may also cause the fixture parts to vibrate. In this case, tighten all screws and wedge the louvers or glass in place. If the noise persists, disconnect the circuit and test the ballast for a short circuit between any sets of contacts.

6. If a fixture constantly flickers on starting in a cold location, such as a garage, there may be no defect. Fluorescent lamps will only start properly when the temperature is above 50°F (10°C). To eliminate this problem, buy special low-temperature lamps and ballasts and install them in place of the standard unit.

Electrical Motors

Nearly all the motors used in a household generate less than 1 hp, often much less; also, most work on 115 V and are single-phase induction motors. You do not have to be an expert electrician to find the trouble in these simple mechanisms and nine times out of ten you can correct it with simple tools and materials.

If a motor does not start, check to see if it is getting power. This seems rather obvious but it is the most common cause of trouble. Is the plug in securely? Is it in at all? Is the circuit live? Check the outlet receptacle with a test lamp.

If the outlet is live, check that current is getting to the motor. Use your test lamp again and carefully connect the probes to the input terminals of the motor. If the lamp still lights, the line cord and plug are working properly and the trouble is within the motor. Unplug the line cord to disconnect all power to the motor; tests from now on will require the VOM or similar tester that has its own battery-powered supply. If the circuit includes a switch, check the switch terminals for loose connections that can be repaired. Then check across the switch terminals for open circuits when the switch is off and continuity when the switch is on. If the switch is bad, replace it rather than attempting any repair.

If you must go deeper into the motor to find the trouble, proceed cautiously. Use numbered strips of tape to mark each wire as you remove it; never depend on your memory to help you reassemble it. When removing end bells (covers on each end of the motor cylinder) from a motor, mark the body of the motor housing and the end bells in some way that you can be sure of replacing the end bells in

Dab paint across
both housing to
end bell seams

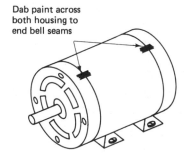

Fig. 12-5. Marking motor end bells before disassembly to assure proper alignment for reassembly.

exactly the same position as they were before disassembly. This can be done with paint as shown in Figure 12-5.

Disconnect the insulated terminal plate, if one is attached to an end bell, and push it inside the housing. Mark the wires and terminals with numbered strips of tape and make a sketch of just how the wires are attached before disconnecting any wires. Remove the nuts from the bolts that hold the end bells. Slip off one end bell (the one with the terminal board would be the best choice). If the nuts and through bolts are removed and the end bell still cannot be taken off, use a small cold chisel and hammer to separate the end bell and housing. Too much leverage at one point could damage the bearings, so tap gently around the whole circumference to work the end bell free. Then remove the core of the motor from inside the housing. Note the location of washers on the shaft when you pull the core free.

Removing the end bell may have exposed a pair of brushes (actually slugs that look like they are made of pencil lead material) and a segmented commutator. Check the wires or pigtails leading to the brush holders (see Fig. 12-6). One of these could have worked loose from its connections or worn bare. Press the brushes into their holder and see that they spring out freely. They may be stuck in the holder so that the spring behind the brush cannot press the brush against the commutator. Reconnect any loose brush wires and wrap tape around any wire whose insulation may have worn through. Replace both holder springs if either is weak—they are generally set loosely in the holders under the plunger-like brushes or they may bear on the outside of a lever assembly to which the brush is fastened. Take an old spring along if you want to shop for a replacement. Also, if the brushes look badly worn or damaged in any way, replace them.

Check the commutator. The commutator is the segmented, cylindrical metal section mounted at the end of the rotor shaft and against which the brushes bear. If the brushes checked out okay, look for a loose connection where the rotor

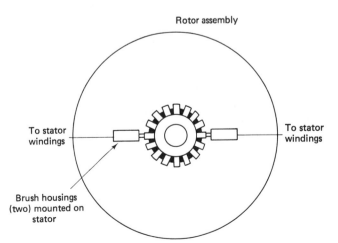

Fig. 12-6. Check the brushes and commutator, which will be exposed after the bell has been removed. Be sure the wires leading to the bush holders are well connected and bushes contact the commutator.

Rotor assembly

To stator windings

To stator windings

Brush housings (two) mounted on stator

windings are soldered to the commutator segments. If the motor has stopped with one of its brushes at such a segment, it will refuse to start again. In other positions, it will turn over but with poor power and possibly sparking may result. If you find a loose connection and you are handy with a soldering iron, resolder the wire to the corresponding commutator segment. Take care not to touch the coils with the hot iron. Then press any slack against the body of the coil and brush on enough shellac to hold it in place. Do not run the motor until the shellac has dried. If you do not want to try soldering, a motor repair shop will probably handle the job from this point at a much lower charge than you would normally pay for a complete repair.

If all is well up to this point, check for defective stator connections next. The stator winding is fixed to the inside of the housing. If a winding has come loose from a terminal, fasten it back. When a broken wire is discovered, splice the ends together (see Fig. 5-4) and cover with tape.

You may also come across motors that have no brushes or commutator. This type is called a shaded-pole motor and is easily recognized by two heavy loops of bare copper wire fitted into slots. These motors operate fairly silently. The stator connection check is the only inspection these motors require; they have only two wires running from their terminal posts.

If the above checks fail to uncover the problem, there is a break or burn-out inside one of the motor windings. You could confirm this with a VOM check for shorts between any two sets of rotor or stator windings. In this case, you had better leave it to a professional repair shop for repair or replace the entire motor.

Sometimes a motor will hum but still fail to turn over. This can be caused by foreign matter wedged between the rotor and stator; clearance between them is very small. Shut off current quickly and try to turn the shaft by hand. If you suspect that a mechanical block is the problem, disassemble the motor, clean the rotor and stator outside and inside diameters, and reassemble for recheck.

If the motor still balks, you may be faced with bearing problems; inspect the bearings in the end bells carefully for excessive wear, damage, or lack of lubrication. Loosen the end bells and try shifting them to be sure that bearings align properly. Try snugging up the bolts. A loose end bell could allow misalignment.

If a motor has no brushes or commutators, but is not the shaded-pole type just covered, it will have four wires leading from its terminals. One pair leads to the main or running coils of the stator and the other pair to the starter coils through a centrifugal switch. The switch is called a centrifugal switch because it operates (opens and closes) under the influence of centrifugal force as the rotor spins. Spring-held weights keep contacts closed until the rotor speeds up. In the starting position, with the contacts held closed, the current flows through the starter windings. Then, as the motor revs up, the spring-held weights fly out by centrifugal force and a lever arrangement moves a ring axially along the shaft, causing the contact to break. Gum on the shaft or damage of some sort could hold the contact apart at all times. Check that contacts can make and break

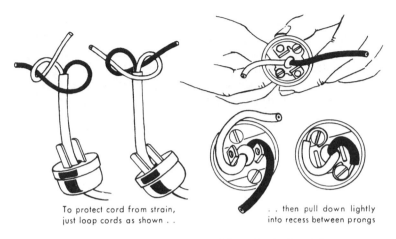

To protect cord from strain, just loop cords as shown . .

. . then pull down lightly into recess between prongs

Fig. 12-7. Underwriter's knot. Courtesy of Sears, Roebuck and Company.

smoothly. Check across the contacts with your VOM to see that contact resistance is not excessive. Sandpaper contact surfaces lightly if they are dirty.

Plug, Cord and Lamp Repair

Most work on damaged plugs, cords, and lamps is so simple and straightforward that we need hardly mention it. However, there are some parts we could profitably cover.

In Chapters 4 and 5, we didn't describe lamp or plug hardware and extension cords in detail or special methods for handling them. A quick survey might now be helpful. Among the plugs available you will find three-blade grounding plugs only for use in 3-blade grounded receptacles, 230-watt plugs for large appliances and quick connect plugs which snap onto lead wires and make contact with the wire by sharp points which cut right through the outer insulation. Another plug you should use is a plug with a narrow neck and round body that allows room for an Underwriter's knot. The Underwriter's knot is simply a strain relief mechanism for the cord. No matter how often we are reminded to pull on the plug, not the cord, people will still jerk the cord. This special knot (see Figure 12-7) provides some protection from damage.

About the three-wire grounding plug, we might just state the obvious: the line or extension cord must contain hot, neutral and ground wires, and the three-wires must be connected by screw fastening to the indicated terminals of the plug. The 230-volt plug has blades arranged in a unique shape to plug into 230-volt receptacles only. There are several characteristic shapes for the blades, depending on the amperage to be drawn. The line or power cord for this plug would also have to contain conductors sized for the same amperage the plug is designed to carry.

Very old cords, especially those used near heaters of any sort, should be checked periodically for cracking of the insulation. If any cracks are found, don't hesitate; replace the cords immediately to prevent more serious trouble if the insulation breaks away and causes a short circuit. In all cords, the conductors are made of many strands of fine wire to provide flexibility. A cotton covering over the twisted strands helps retain flexibility by keeping the insulation from bonding to the wires. The insulation material and its thickness vary according to the use for which the cord is designed.

Underwriter's Cord Type SPT is commonly used for lamps and has conductors which are insulated by plastic. Type SP is similar, but has rubber insulation. The two wires are covered by jute and a tough outer coat of rubber insulation is added. When a cord has to tolerate considerable amounts of heat, such as in supply lines to irons, toasters, etc., Code Type HPD is used. A layer of asbestos covers each wire. The wires are then twisted together and cotton is used for a durable outer cover.

When replacing any cord, be sure to duplicate the discarded cord exactly, if possible. Also match the length as nearly as possible, because a cord that is too long could have enough additional resistance to affect performance of the appliance. Figure 12-8 shows how to select an extension cord and relates amperage rating to maximum cord length.

Be sure you use flexible extension cord correctly. The NEC prohibits its use for fixed wiring in a structure in which it will run through holes in walls, ceilings or floors, through doorways or windows, or be attached to building surfaces. Those insulated staples you see for sale (and may have even used from time to time) are strictly forbidden for flexible extension cords carrying normal house voltage.

Damaged cords may be temporarily repaired by cutting out the damaged section and splicing and taping the remaining wires together. However, we think this is just asking for trouble and should only be used for emergency repairs. You should replace the cord as soon as possible. The same is true for lamp sockets that

CORD	AMPERE RATING FOR 110–120 V.A.C. TOOLS															
LENGTH	0-5A.	6A.	7A.	8A.	9A.	10A.	11A.	12A.	13A.	14A.	15A.	16A.	17A.	18A.	19A.	
25 FT.	18	18	18	18	18	18	16	16	16	14	14	14	14	14	12	12
50 FT.	18	18	18	18	18	18	16	16	16	14	14	14	14	14	12	12
75 FT.	18	18	18	18	16	16	16	16	16	14	14	14	14	14	12	12
100 FT.	18	16	16	16	16	16	16	16	14	14	14	14	14	14	12	12
125 FT.	16	16	16	16	16	14	14	14	14	14	14	12	12	12	12	12
150 FT.	16	16	14	14	14	14	14	14	14	12	12	12	12	12	12	12

*Wire sizes are AWG (American Wire Gage) Recommendations are minimum allowable.

Fig. 12-8. Table relating amperage rating to maximum cord length.

Fig. 12-9. To replace a lamp socket, unplug the lamp, remove the lamp shade, shade holder and bulb, loosen the nut under the base of the lamp and loosen a set screw at the bottom of the socket. The socket can then be unscrewed from the lamp table that carries the wires. Pull the socket to draw several inches of wire from the top of the lamp and allow room to work on the socket.

Pull the socket shell apart as shown in (A) above and disconnect the wires from terminal screws as shown in (B). Remove and discard the socket and replace it with a new socket. Note the wire ends in (B). For safety's sake, you should split the conductor insulation for a few inches, then make an underwriter's knot (Figure 12-6) in the conductors before connecting the leads to the terminal. This knot will jam in the bottom of the socket so that a pull on the wires will not pull the leads off the terminal screws. (Courtesy of Leviton Manufacturing, Inc.)

contain the lamp's ON-OFF switch. Replace the socket as shown in Figure 12-9. Again, the lamp must be unplugged before you start this procedure.

There is no need for a detailed discussion of how to replace a wall switch or dual receptacle outlet. We discuss their installation in Chapters 7 and 8 and the same procedures should be followed for their replacement. Just remember to turn off the primary input power.

Appliance Repair

When repairing appliances, your first source of information should be the instruction sheet or booklet that came with the appliance. These are sometimes quite sketchy, but even a little data is better than nothing. Hopefully you always make a habit of saving these sheets whenever you make a purchase.

In many instances, repairing an appliance yourself is much more economical than replacing the appliance with a new one. A toaster's delicate heater ribbon could easily break if the toaster drops. Should you replace the toaster for probably more than initial cost, or should you replace the heater element for a few bucks? Depending on the brand, the replacement might be made in as little as 20 minutes. Worn motor brushes might cause your vacuum cleaner to break down. It takes just a few minutes and pennies to save the expensive unit by replacing the worn brushes.

If the appliance operates but does not perform well, check carefully how it is operating and listen for unusual sounds. These easy steps may help you to determine whether the defect is mechanical or electrical.

You will probably have to disassemble the appliance to some extent before you can test further. Look closely for the clips and screws that hold the appliance together. Generally they are hidden so that they do not spoil the appearance of the appliance. Many times the fastening is made with a small metal tab inserted in a slot and bent over. Also, plastic parts with tabs are often held by a spring action and you will have to pry with a screwdriver or squeeze with your fingers to release the tab.

Some appliances have switch controls or heater elements or both. Many appliances contain wires joined by lug terminals crimped on the end, as shown in Figures 12-10 and 12-11. This requires purchase of a special kit that has both the terminals and a crimp tool, but it is relatively inexpensive and provides a very effective connection method with which you should be familiar.

Testing of the plug, cord, switch, and load has already been described. Most of it can be done using the RX 1 scale of the VOM, because accurate readings within a few ohms are not required. More often than not, the major load or one lead of it through one portion of the appliance will fail to open or short, rather than having a small shift of resistance. The VOM can therefore sort out whether it is the switching and control circuits, the resistance elements, or the motor that has failed.

Bimetal Switches and Thermostats

One particularly interesting element that you will encounter in small and large appliances is the bimetal switch or thermostat. The bimetal switch is the heart of an appliance thermostat (like your heating system unit) and is used to switch resistance heating coils on and off to maintain a desired temperature. Irons,

Fig. 12-10. With wire in the proper jaw slot of this tool, only the insulation is cut when the tool closes. Insulation can then be easily slipped off the end of the conductor. (Graf–Whalen photo.)

toasters, coffee makers, waffle irons, and other electric cooking utensils all use thermostats. So do gas and electric clothes dryers. In this type unit, two metal strips made of different metals that expand at different rates when heated are fused (welded or riveted) together lengthwise. An electrical contact point is placed at the one end of the joined strips and the other end is firmly fastened to the thermostat housing. When the strip is heated, one metal begins to expand faster than the other. Because they are fused together, there cannot be any motion between them lengthwise and therefore the arm bends.

In operation, the heated bimetallic arm bends until its contact no longer touches the other, fixed contact of the switch. This breaks the circuit to the heating element. When the arm cools and straightens, the contacts touch again. The circuit is then closed again and reheating occurs. This cycle can be repeated continuously with as little as 5° to 10° temperature variation. One contact of the thermostat is adjustable (just as your furnace thermostat) and can be moved in and out by turning the temperature dial.

Major appliances such as washers, dryers, and ranges also make use of still other controls with which you must become familiar. Here again, your manufacturer's instructions will be very important. One such control is called a relay (see also Chapter 11), which is really an electrically-operated remote switch. It permits current flowing in one circuit to actuate contacts that start or stop current in a second circuit. Relays work electromagnetically, as described in Chapter 1, and generally provide for a small current flow to control the flow of a larger current. A rod with a coil of wire around it works as an electromagnet. When energized with current, the electromagnet attracts a movable metal arm containing an electrical contact. This contact can therefore be moved by an electrical signal from a remote point to open or close a circuit.

A good example of a relay application is the gas safety relay used in a clothes dryer. Here, the relay's contacts control power to a solenoid valve in the gas line. The relay coil is energized through a bimetal switch that senses the heat of gas ignition in the combustion chamber. The relay closes when combustion takes place, allowing the valve to admit gas from the supply line. If combustion ceases, the bimetal straightens, the relay opens, and the valve slams closed, preventing gas from entering the appliance. A failure in any of these will render the dryer inoperative, so each item must be tested.

Relays on appliances such as washers and dryers work this way and allow small currents to control the flow of large currents to motors, and so on. Dirty

Fig. 12-11. After wire ends are stripped, as shown in Fig. 12-10, terminal connectors may be crimped onto the bare conductor ends, or several wires may be spliced end-to-end or in pigtail splice by the same tool. Crimp connectors are permanently bonded to the wire and cannot be removed. Courtesy of Sears, Roebuck and Company.

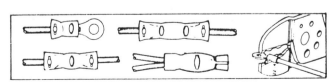

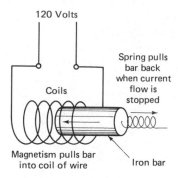

120 Volts

Coils

Spring pulls
bar back
when current
flow is
stopped

Magnetism pulls bar
into coil of wire   Iron bar

Fig. 12-12. How a solenoid works.

contacts or broken wires in the coils can cause a breakdown. If you suspect a relay is causing your trouble, perform a continuity check of the coil and the relay contact circuit. Do not try any repairs on these units. If you find that one is defective, replace it with a new unit. When you remove the relay, or any other component for that matter, tag all wires before you disconnect them to be sure that you can install the new one easily.

The solenoid mentioned previously (see Figure 12-12) also uses the electromagnetic effect, but instead of just moving a contact arm, the solenoid can perform mechanical work by pushing or pulling a soft iron bar in the coil. Gear shift levers, drive-belt idler pulleys, water and gas valves, and similar components can all be operated by this mechanism. Again, you can check these devices for continuity of the coil windings, but you should also check to see that parts move freely and are not binding. Sometimes a little cleaning is all that is needed. Solenoids are rarely lubricated because it can gum-up their operation.

Although there are so many brands of washers and dryers, most of them can be diagnosed from the instructions. All washers have a motor, drive belt, gears, pump, and timer in a large metal housing. Dryers use tumbling drums, electric heaters, and air blowers. Thermostats and timers control the action and cycle the dryer's heat and tumbling as required. With either of these major units, your VOM will be your best aid. After unplugging or disconnecting the appliance from the branch circuit or setting the branch circuit breaker off, make continuity checks along every circuit that you can identify and check for breaks or binding. Physical examination in conjunction with a few electrical checks is remarkably effective in locating the source of a problem.

Electric ranges may seem complicated, but the circuits turn out to be quite simple. They generally operate on 115/230-V three-wire inputs, with voltages switched in to the resistance heating elements, as shown in Figure 12-13. Each surface unit comprises both an inner and outer element and, by the switching method shown, seven different heat levels can be obtained. These inputs are switch-controlled, manually, for each heater. Switches as well as heater elements can be checked for continuity *with the range circuit disconnected from the power line.* The ovens normally have heater coils in the top surface and bottom surface of the cavity, with a thermostat for heater control. Because ranges have few

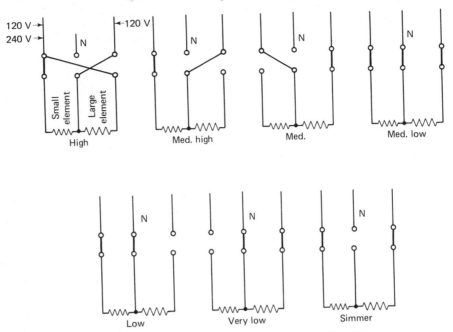

Fig. 12-13. Seven-heat switching of electric range from 120/240-voltline.

moving parts, thermostats, heater coils, and switches comprise most of the trouble spots you will encounter. Again, with your VOM and manufacturer's instructions, you should be able to cover all but the most complex repair jobs. Do not be afraid to call the repairman as a last resort, but let's hope that with the instructions here and a bit of practice, these last resorts will become fewer and fewer.

GLOSSARY

ac (Alternating current):   An electric current that reverses its polarity, hence the direction of current flow, at regularly recurring intervals. Each complete polarity reversal is called a *cycle*. The number of cycles occurring in a second is called the *frequency,* expressed in Hertz, or cycles per second.

Ampacity: Current-carrying capacity of a conductor, expressed in amperes. For a given gauge (size of conductor), the ampacity depends on the material of the conductor (e.g., copper can carry more amperes of current than aluminum for a given conductor size).

Ampere: A measure of the amount of electrical current flow. One ampere is approximately 6.24×10^{18} electrons flowing past a point in a circuit in exactly one second.

Apparent Power: The product of volts times amperes in an a-c circuit. It is not always equal to actual, effective power because circuit reactance throws voltage and current out of step. Remember that reactance stores some of the energy; hence, this energy is put to work in the device.

Approved: Signifies that the minimum standards established by a cognizant authority have been met.

Armored Cable: A flexible, metallic-sheathed cable used for interior wiring. A commercial tradename for this cable is BX.

Average Voltage (used also for average current):   Because a-c voltage and current vary in a wavelike manner, the average voltage applied to a circuit is only equal to 63.6% of the maximum or peak voltage.

Ballast: A magnetic coil that adjusts current through a fluorescent tube, providing the necessary current surge to start the lamp. It also maintains even current for continuous operation.

Bonding: The permanent joining of metallic parts to form an electrically conductive path that will assure electrical continuity and capacity to conduct safely any probable fault current. (See also *Grounds*.)

Box: An electrical wiring device, required in most electrical safety codes, that is used to house wire terminations at points where they connect to other wires (see *Junction Box*), switches, outlet plugs, and the like. (See *Device*.)

Branch Circuit: A circuit that supplies a number of outlets for lighting or appliances.

Bus Bar: A heavy conductor at the service entrance to which branch circuits are connected.

BX: Commercial tradename for a type of flexible, armored electrical cable.

Cable: A conductor assembly consisting of two or more wires insulated from each other and grouped together in an overall covering. (See also *BK* and *Romex*.)

Charge: See *Electrical Charge*.

Circuit: A complete, closed path of electrical conductors, leading from a source (generator or battery), through components such as motors or lamps, and back to the opposite terminal of the source.

Circuit Breaker: An electro-magnetic or thermal device that opens a circuit when the current exceeds a predetermined amount. Unlike fuses, a circuit breaker can be reset. (See *Fuse*.)

Conductor: A low-resistance material through which electricity flows easily. Copper wire, used in most house wiring, is a good conductor.

Conduit: Metal or fiber pipe or tube used to enclose electrical conductors.

Continuity: An uninterrupted electrical path. A continuity tester (test light) connected to two points of a circuit or wire is used to determine whether there are any hidden breaks or poor connections that introduce resistance.

Current: The movement or flow of electrons. Also, the time rate of electron flow, measured in amperes. (See *Amperes*.)

Cycle: One complete positive and negative alternation of a current or voltage.

dc: Abbreviation for direct current.

Device: In accordance with the National Electrical Code, it is a component of an electrical system that does not normally carry current, such as a junction box.

Direct Current: An electric current that flows in only one direction. It has a fixed polarity, designated by ($+$) and ($-$).

Electrical Charge: The electrical energy of a body or particle. The electron has an inherent negative charge, the proton has an inherent positive charge.

Electromagnet: A magnet made by passing current through a coiled wire conductor that surrounds a soft iron core.

Electromagnetism: The relationship between electricity and magnetism. By means of this phenomenon, either form of energy may be converted to the other.

Electromotive Force: Can be thought of as the force or pressure that makes electric current flow in a circuit. (See *Voltage*.)

Electron: The negatively charged particle of an atom. Electric currents comprise the flow or movement of electrons in a conductor.

Feeder: The circuit conductors between the service equipment and the branch circuit overcurrent device.

Field: The volume of space around a magnet or electrical charge permeated by invisible lines of force (magnetic or electric).

Fish Tape: Flat, steel spring wire with hooked ends that is used to pull wires through enclosures such as conduits or walls. "Fish" refers to the technique of hooking two tapes together in a blind location so that wire may be connected to the tape and pulled through.

Fuse: A protective device inserted in series with a circuit. It contains a soft metal link that will melt or break when current exceeds a rated value for a specific time period, breaking the circuit and turning off the current.

Generator: A machine containing an electrical device that produces an electrical energy output from an input of mechanical energy. Generators can receive mechanical energy from steam engines, nuclear reaction, water-powered turbines, gasoline engines, etc.

Ground: A connection between an electrical circuit and the earth (or a common body that serves as a reference point in place of the earth). In a house circuit, the service entrance neutral wire is connected directly to earth, and outlet boxes and other elements are bonded to the service entrance ground.

Grounding Conductor: The one wire in a cable that normally carries no current. Instead, it connects exposed metal surfaces to the service entrance ground to prevent hazard in case of insulation breakdown between the current-carrying wires and the exposed surfaces. If a hot wire touches an outlet box, wallplate, or the frame of an appliance, the grounding wire provides a short circuit back to the ground neutral connection at the service entrance to activate overcurrent protection and terminate a fire or shock hazard.

Hanger: A metal or insulated strap used to support electrical cable at intervals between one point of connection and another.

Hertz: A unit of frequency equal to one cycle per second. Named in honor of Heinrich Hertz, an early electrical experimenter. Replaces cps in frequency designation (e.g. 60 Hz equals 60 cps).

Horsepower: A unit of power. Its electrical equivalent is 746 W.

Hot Wire: A common term referring to the wire of a house circuit that is not connected to ground and is therefore at high voltage with respect to ground or neutral.

Impedance: The total opposition to alternating current created by an electrical circuit. It may consist of any combination of resistance, inductive reactance and capacitive reactance. (See *Resistance* and *Reactance*.)

Induced Current: Current in a conductor caused by a magnetic field fluctuating around and about the conductor.

Junction Box: A box in which several conductors are connected together, such as for branching a circuit into several parallel paths.

National Electrical Code: Rules and guidelines for safe use and installation of electrical material and equipment. It is published by the National Fire Protection Association to promote safe methods and prevent electrical hazards. The NEC is not a book of instructions nor is it a law. However, many municipalities use the code as the basis for local wiring and inspection regulations. You may order a copy from the National Fire Protection Association, Publications Sales Dept., 470 Atlantic Avenue, Boston, Mass., 02210.

Negative Charge: The electrical charge inherent in an electron or carried by a body that has an excess of electrons.

Neutral Wire: The wire in a cable that is maintained at zero voltage. It is the return path to the source. Hence, all current flowing through the hot wire must also flow through the neutral wire.

Ohm: The unit of electrical resistance. One ohm is equal to the resistance of a conductor in which one volt of potential difference will produce a current flow of one ampere.

Open Circuit: An electrical circuit through which no current can flow because there is a physical break in the path, such as is caused by opening a switch, disconnecting a wire, or burning out a lamp filament or fuse.

Outlet: A box, metallic or fiber, in which electrical wiring is connected to electrical components, such as a convenience plug. (See *Device* and *Receptacle*.)

Overcurrent Protection Device: A component, such as a fuse or circuit breaker, designed to open an electric circuit automatically if the current exceeds a predetermined safe value.

Overload: Current demand greater than that for which a device or circuit is rated.

Parallel Circuit: A circuit that provides more than one path for electrical current to flow.

Polarity: A term used in d-c circuits to identify whether a point is positive or negative in charge.

Polarized Plug: A plug with blades physically designed so that they may enter a receptacle in only one orientation.

Potential Difference: The amount of electrical pressure at one point in a circuit with respect to another point. It is usually developed by connecting the source's high and neutral sides across one or more impedances in series. The potential difference across any of the impedances is measured in volts and is also called voltage drop. (See *Voltage* and *Voltage Drop*.)

Power: The rate of doing work or the rate of expending energy. The unit of electrical power is the watt. (See *Watt*.)

Primary: The winding of a transformer or similar device that receives the input power. (See also *Secondary*.)

Raceway: A mechanical, surface-mounted channel designed to support wires or cables in a path from one connection point to another.

Reactance: A circuit's opposition to the flow of ac, caused by either inductance or capacitance.

> **1.** Inductance: The storage property of magnetic circuit elements, usually coiled conductors, by which a change in electric current produces, by electromagnetic induction, an electromotive force or voltage opposite to the voltage causing the current to flow. Called inductive reactance, this opposing force increases with increasing frequency of a-c current.

> **2.** Capacitance: The property of circuit elements, called capacitors, that permits the storage of electrons on conductive plates separated by an insulator. As electrons build up on one plate, the current flow decreases due to repulsion between similar electric charges. Called capacitive reactance, this opposing force decreases with the increasing frequency of a-c current.

Receptacle: A housing that contains separated hot and neutral contacts and that is installed at an outlet to supply current to a single extension cord plug. It may also supply a ground connection. The standard practice is to mount two receptacles in one body, connected to common hot and neutral wires [a duplex (or dual) receptacle].

Relay: An electrical component that opens or closes contacts, thus opening and closing circuits, by means of remote-control current application to an electromechanical element.

Resistance: The property of a conductor that resists (opposes) the passage of electricity because of the conductor's atomic nature. Electrons crowding through the conductor cause the atoms to vibrate. This vibration shows up as heat and uses up part of the electrical energy. The unit of measure of resistance is the *ohm*. It is also called ohmic resistance. (See *Ohm*.)

Romex: Commercial tradename for nonmetallic-sheathed electrical cable that is used for indoor wiring.

Secondary: The winding in a transformer or similar device in which a current is induced by a primary winding; the secondary winding is the output side. (See also *Primary*.)

Series Circuit: An electrical circuit in which there is only a single path for electric current to flow. Interrupting the path at any point causes cessation of current.

Short Circuit: A condition, usually unintentional, in which a low resistance path develops between two points of different potential in a circuit. It results in a large current flow and is often accompanied by sparks and a fire hazard. Overcurrent devices (fuses and circuit breakers) are provided to mitigate these faults.

60-Cycle Current: ac that completes 60 cycles within one second. Also called 60 Hz 60 Hz is standard frequency for North America. Several other countries use 50 Hz (notably in Europe).

Solder: A metallic alloy of tin and lead that has a low melting point and can be used to join bare copper conductors. When melted to flow around mechanically joined conductors, it forms a good electrical and mechanical alloy bond upon cooling and hardening.

Solderless Connector: A mechanical device, usually insulated by plastic, which can be finger-fastened over the exposed and joined ends of several wires to provide a firm connection between the wires. They are also called wire nuts.

Solenoid: An electromagnetic coiled wire element in which electric current through the coiled conductor causes a plunger (technically, an *armature*) to move along the axis of the coil. The stroke of the solenoid is sufficient to physically move valves, contacts, and other devices in response to the electrical input to the coil.

Split Receptacle: A dual receptacle in which each of the two receptacles is connected to a different branch circuit, not to a common circuit.

Starter: An automatic switch that opens to trigger the ballast that causes current to flow through a fluorescent tube. Thereafter, it stays open as long as current is flowing through the fluorescent lamp and the circuit is completed.

Switch: A component that is used to complete (close) or break (open) an electrical circuit or to divert electric current from one circuit to another. Switches are used only in hot wires, never in neutral wires.

Transformer: A device composed of two or more coils of electrical conductors that are linked only by means of magnetic lines of force. Usually both coils are wound on a core of magnetic material, such as silicon steel. They are used to increase or decrease voltage and to transfer energy from one circuit to another through a magnetic pathway.

Underwriter's Knot: A knot used to tie two insulated conductors at the terminals inside an electric plug. The knot bears against the physical body of the plug. It is used to relieve strain on the terminal connection.

UL Label: It is a label applied to manufactured devices that have been tested for safety by the Underwriter's Laboratory and listed within a given class for marketing. The Underwriter's Laboratory is a nonprofit institution supported by insurance companies, manufacturers, and other parties interested in electrical safety. UL does not approve devices. Rather, it attests to the fact that a specific item has passed minimum safety standards for listing within a particular class.

Voltage: The electromotive force or potential difference between two points of a circuit that causes electric current to flow. It is measured in volts. One volt creates a current of one ampere through a resistance of one ohm.

Volt-Ampere: In an a-c circuit, a unit of measurement of electrical power equal to the product of volts times amperes. In dc, one volt-ampere equals one watt and in a c it is a unit of apparent power. (See *Apparent Power.*)

Voltage Drop: A drop in line voltage through conductors and circuit components caused by their resistance to current flow. In a-c circuits, the resistance is called impedance. (See *Impedance.*)

Watts: A unit of measurement of electric power. In a d-c circuit (or an a-c circuit with pure resistive load), the product of volts times amperes equals the watts of electricity used. 1000 watts used for one hour equals *one kilowatt hour* and is the standard method of measurement that utility companies use for billing.

Wire Gauge: A standard numerical method of specifying the physical size of a conductor. The American Wire Gage (AWG) series is the system most commonly used.

INDEX